Communication Networks
A Concise Introduction
Second Edition

Praise for *Communication Networks: A Concise Introduction*

This book is a welcome addition to the Networking literature. It presents a comprehensive set of topics from an innovative and modern viewpoint that observes the rapid development that the field has undergone. It is informed, informative, and useful to students, faculty, and practitioners in the field. This book is a must have!

-Anthony Ephremides, *University of Maryland*

Computer networks are extremely complex systems, and university-level textbooks often contain lengthy descriptions of them that sacrifice basic conceptual understanding in favor of detailed operational explanations. Walrand and Parekh depart from this approach, offering a concise and refreshing treatment that focuses on the fundamental principles that students can (and should) carry forward in their careers. The new edition has been updated to cover the latest developments and is of great value for teaching a first upper-division course on computer networks.

-Massimo Franceschetti, *University of California, San Diego*

The book presents the most important principles of communication network design, with emphasis on the Internet, in a crisp and clear fashion. Coverage extends from physical layer issues through the key distributed applications. The book will be a valuable resource to students, instructors, and practitioners for years to come.

-Bruce Hajek, *University of Illinois at Urbana-Champaign*

Conceptual clarity, simplicity of explanation, and brevity are the soul of this book. It covers a very broad swathe of contemporary topics, distills complex systems into their absolutely basic constituent ideas, and explains each idea clearly and succinctly. It is a role model of what a classroom text should be. I wish there had been such a book when I was learning about communication networks.

-P. R. Kumar, *Texas A&M University*

This book focuses on the basic principles that underlie the design and operation of the Internet. It provides a holistic account of this critical yet complex infrastructure and explains the essential ideas clearly and precisely without being obscured by unessential implementation or analytical details. It is the best introduction to networking from which more specialized treatment of various topics can be pursued.

-Steven Low, *California Institute of Technology (Caltech)*

Communication Networks: A Concise Introduction by Jean Walrand and Shyam Parekh is an amazing book. Jean and Shyam are in the unique position for writing this book because of the foundational contributions they made to the area and their many years of teaching this course at UC Berkeley. This book covers many important topics ranging from the architecture of the Internet, to today's wireless technologies, and to emerging topics such as SDN and IoT. For each topic, the book focuses on the key principles and core concepts, and presents a concise discussion of how these principles are essential to scalable and robust communication networks. Mathematical tools such as Markov chains and graph theory are introduced at a level that is easily understandable but also adequate for modeling and analyzing the key components of communication networks. The comprehensive coverage of core concepts of communication networks and the intuition/principle-driven approach make this book the best textbook for an introductory course in communication networks for those students who are interested in pursuing research in this field. It is certainly a must-have book for students and researchers in the field.

-Lei Ying, *Arizona State University*

Synthesis Lectures on Communication Networks

Editor
R. Srikant, *University of Illinois at Urbana-Champaign*

Founding Editor Emeritus
Jean Walrand, *University of California, Berkeley*

Synthesis Lectures on Communication Networks is an ongoing series of 50- to 100-page publications on topics on the design, implementation, and management of communication networks. Each lecture is a self-contained presentation of one topic by a leading expert. The topics range from algorithms to hardware implementations and cover a broad spectrum of issues from security to multiple-access protocols. The series addresses technologies from sensor networks to reconfigurable optical networks.
The series is designed to:

- Provide the best available presentations of important aspects of communication networks.

- Help engineers and advanced students keep up with recent developments in a rapidly evolving technology.

- Facilitate the development of courses in this field

Communication Networks: A Concise Introduction, Second Edition
Jean Walrand and Shyam Parekh
2017

BATS Codes: Theory and Practice
Shenghao Yang and Raymond W. Yeung
2017

Analytical Methods for Network Congestion Control
Steven H. Low
2017

Advances in Multi-Channel Resource Allocation: Throughput, Delay, and Complexity
Bo Ji, Xiaojun Lin, Ness B. Shroff
2016

Communication Networks: A Concise Introduction, Second Edition

Jean Walrand and Shyam Parekh

ISBN:978-3-031-79280-9   paperback
ISBN:978-3-031-79281-6   ebook
ISBN:978-3-031-79283-0   epub
ISBN:978-3-031-79282-3   hardcover

DOI 10.1007/978-3-031-79281-6

A Publication in the Springer series
SYNTHESIS LECTURES ON COMMUNICATION NETWORKS

Lecture #20
Series Editor: R. Srikant, *University of Illinois at Urbana-Champaign*
Founding Editor Emeritus: Jean Walrand, *University of California, Berkeley*
Series ISSN
Print 1935-4185   Electronic 1935-4193

Communication Networks
A Concise Introduction
Second Edition

Jean Walrand and Shyam Parekh
University of California, Berkeley

SYNTHESIS LECTURES ON COMMUNICATION NETWORKS #20

ABSTRACT TO THE SECOND EDITION

This book results from many years of teaching an upper division course on communication networks in the EECS department at the University of California, Berkeley. It is motivated by the perceived need for an easily accessible textbook that puts emphasis on the core concepts behind current and next generation networks. After an overview of how today's Internet works and a discussion of the main principles behind its architecture, we discuss the key ideas behind Ethernet, WiFi networks, routing, internetworking, and TCP. To make the book as self-contained as possible, brief discussions of probability and Markov chain concepts are included in the appendices. This is followed by a brief discussion of mathematical models that provide insight into the operations of network protocols. Next, the main ideas behind the new generation of wireless networks based on LTE, and the notion of QoS are presented. A concise discussion of the physical layer technologies underlying various networks is also included. Finally, a sampling of topics is presented that may have significant influence on the future evolution of networks, including overlay networks like content delivery and peer-to-peer networks, sensor networks, distributed algorithms, Byzantine agreement, source compression, SDN and NFV, and Internet of Things.

KEYWORDS

Internet, Ethernet, WiFi, Routing, Bellman-Ford algorithm, Dijkstra algorithm, TCP, Congestion Control, Flow Control, QoS, LTE, Peer-to-Peer Networks, SDN, NFV, IoT

Contents

Preface

These lecture notes are based on an upper division course on communication networks that the authors have taught in the Department of Electrical Engineering and Computer Sciences of the University of California at Berkeley.

Over the thirty years that we have taught this course, networks have evolved from the early Arpanet and experimental versions of Ethernet to a global Internet with broadband wireless access and new applications from social networks to sensor networks.

We have used many textbooks over these years. The goal of this book is to be more faithful to the actual material we present. In a one-semester course, it is not possible to cover an 800-page book. Instead, in the course and in these notes we focus on the key principles that we believe the students should understand. We want the course to show the forest as much as the trees. Networking technology keeps on evolving. Our students will not be asked to re-invent TCP/IP. They need a conceptual understanding to continue inventing the future.

Besides correcting the known errors and adding some clarifications, the main changes in this second edition are as follows. Chapter 4 on WiFi has been updated to cover recent advances. Chapter 7 on Transport includes a discussion of alternative congestion control schemes. Chapter 8 on Models has been expanded with sections on graphs and queues. Furthermore, the chapter now explains the formulation of TCP and of sharing a wireless link as optimization problems. Chapter 9 on LTE now discusses the basics of cellular networks and a further exposition of a number of key aspects of LTE. It also includes presentations of LTE-Advanced and 5G. Discussion of WiMAX has been dropped in light of the overwhelming acceptance of LTE. In the Additional Topics chapter (Chapter 12), we have added the following topics: Switches (including Modular Switches, Switched Crossbars, Fat Trees), SDN and NFV, and IoT.

These lecture notes have an associated website[1] that we plan to use for future updates. In recent years, our Berkeley course has also included a research project where the students apply the fundamental concepts from the course to a wide variety of topics related to networking. Interested readers can also find the extended abstracts from these research projects on this website.

Many colleagues take turns teaching the Berkeley course. This rotation keeps the material fresh and broad in its coverage. It is a pleasure to acknowledge the important contributions to the material presented here of Kevin Fall, Randy Katz, Steve McCanne, Abhay Parekh, Vern Paxson, Sylvia Ratnasamy, Scott Shenker, Ion Stoica, David Tse, and Adam Wolicz. We also thank the many teaching assistants who helped us over the years and the inquisitive Berkeley students who always keep us honest.

---

[1] https://bit.ly/2zPXDL3

We are grateful to reviewers of early drafts of this material. In particular, Assane Gueye, Libin Jiang, Jiwoong Lee, Steven Low, John Musacchio, Jennifer Rexford, and Nikhil Shetty provided useful constructive comments. We thank Mike Morgan of Morgan & Claypool for his encouragement and his help in getting this text reviewed and published.

The first author was supported in part by NSF and by a MURI grant from ARO during the writing of this book.

Most importantly, as always, we are deeply grateful to our families for their unwavering support.

Jean Walrand and Shyam Parekh
October 2017

C H A P T E R   1

The Internet

The Internet grew from a small experiment in the late 1960s to a network that connects about 3.7 billion users (in June 2017) and has become society's main communication system. This phenomenal growth is rooted in the architecture of the Internet that makes it scalable, flexible, and extensible, and provides remarkable economies of scale. In this chapter, we explain how the Internet works.

1.1   BASIC OPERATIONS

The Internet delivers information by first arranging it into packets. This section explains how packets get to their destination, how the network corrects errors and controls congestion.

1.1.1   HOSTS, ROUTERS, LINKS

The Internet consists of hosts and routers attached to each other with links. The *hosts* are sources or sinks of information. The name "hosts" indicates that these devices host the applications that generate or use the information they exchange over the network. Hosts include computers, printers, servers, web cams, etc. The *routers* receive information and forward it to the next router or host. A *link* transports bits between two routers or hosts. Links are either optical, wired (including cable), or wireless. Some links are more complex and involve switches, as we study later. Figure 1.1 shows a few hosts and routers attached with links. The clouds represent other sets of routers and links.

1.1.2   PACKET SWITCHING

The original motivation for the Internet was to build a network that would be robust against attacks on some of its parts. The initial idea was that, should part of the network get disabled, routers would reroute information automatically along alternate paths. This flexible routing is based on *packet switching*. Using packet switching, the network transports bits grouped in *packets*. A packet is a string of bits arranged according to a specified format. An Internet packet contains its source and destination addresses. Figure 1.1 shows a packet with its source address A and destination address B. Switching refers to the selection of the set of links that a packet follows from its source to its destination. Packet switching means that the routers make this selection individually for each packet. In contrast, the telephone network uses *circuit switching* where it selects the set of links only once for a complete telephone conversation and reserves the data rate

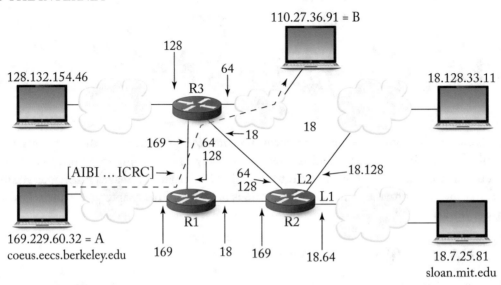

Figure 1.1: Hosts, routers, and links. Each host has a distinct location-based 32-bit IP address. The packet header contains the source and destination addresses and an error checksum. The routers maintain routing tables that specify the output for the longest prefix match of the destination address.

required on that set of links for the duration of the conversation. There is also an intermediate solution using the concept of *virtual circuits* where the set of links that transport the information is also selected once as in circuit switching, but the required data rate is not reserved on those links. Virtual circuits also transport data using packets, and send the packets of a connection along the same set of links. In order to distinguish it from the virtual circuit-based packet switching, the term *datagrams* is often used for the version of packet switching where each router chooses the next link individually for each packet. *Multi-Protocol Label Switching (MPLS)* and *Asynchronous Transfer Mode (ATM)* are examples of virtual circuit-based technologies.

1.1.3   ADDRESSING

In version 4 of the Internet protocol (called IPv4), every computer or other host attached to the Internet has a unique address specified by a 32-bit string called its *IP address*, for Internet Protocol Address. The addresses are conventionally written in the form $a.b.c.d$ where a, b, c, d are the decimal value of the four bytes. For instance, 169.229.60.32 corresponds to the four bytes 10101001.11100101.00111100.00100000. A more recent version of the protocols, called IPv6, uses 128-bit addresses but is compatible with IPv4.

1.1.4   ROUTING

Each router determines the next hop for the packet from the destination address. While advancing toward the destination, within a network under the control of a common administrator, the packets essentially follow the shortest path.[1] The routers regularly compute these shortest paths and record them as *routing tables*.[2] A routing table specifies the next hop for each destination address, as sketched in Figure 1.2.

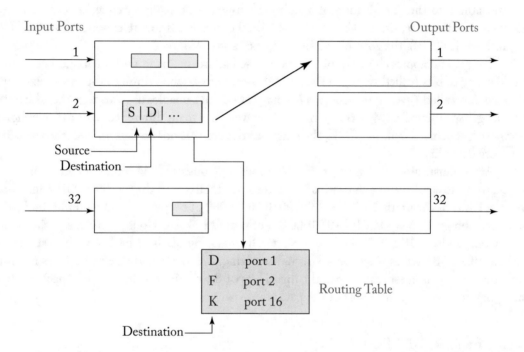

Figure 1.2: The figure sketches a router with 32 input ports (link attachments) and 32 output ports. Packets contain their source and destination addresses. A routing table specifies, for each destination, the corresponding output port for the packet.

To simplify the routing tables, the network administrators assign IP addresses to hosts based on their location. For instance, router $R1$ in Figure 1.1 sends all the packets with a destination address whose first byte has decimal value 18 to router $R2$ and all the packets with a

---

[1]The packets typically go through a set of networks that belong to different organizations. The routers select this set of networks according to rules that we discuss in the Routing chapter.

[2]More precisely, the router consults a *forwarding table* that indicates the *output port* of the packet. However, this distinction is not essential.

destination address whose first byte has decimal value 64 to router *R*3. Instead of having one entry for every possible destination address, a router has one entry for a set of addresses with a common initial bit string, or *prefix*. If one could assign addresses so that all the destinations that share the same initial 5 bits were reachable from the same port of a 32-port router, then the routing table of the router would need only 32 entries of 5 bits: each entry would specify the initial 5 bits that correspond to each port. In practice, the assignment of addresses is not perfectly regular, but it nevertheless reduces considerably the size of routing tables. This arrangement is quite similar to the organization of telephone numbers into [country code, area code, zone, number]. For instance, the number 1 510 642 1529 corresponds to a telephone set in the U.S. (1), in Berkeley (510), the zone of the Berkeley campus (642).

The general approach to exploit location-based addressing is to find the longest common initial string of bits (called *prefix*) in the addresses that are reached through the same next hop. This scheme, called *Classless Interdomain Routing* (or CIDR), enables us to arrange the addresses into subgroups identified by prefixes in a flexible way. The main difference with the telephone numbering scheme is that, in CIDR, the length of the prefix is not predetermined, thus providing more flexibility.

As an illustration of *longest prefix match* routing, consider how router *R*2 in Figure 1.1 selects where to send packets. A destination address that starts with the bits 000100101 matches the first 9 bits of the prefix 18.128 = 00010010'10000000 of output link L2 but only the first 8 bits of the prefix 18.64 = 00010010'01000000 of output link L1. Consequently, a packet with destination address 18.128.33.11 leaves *R*2 via link L2. Similarly, a packet with destination address 18.7.25.81 leaves *R*2 via link L1. Summarizing, the router finds the prefix in its routing table that has the longest match with the initial bits of the destination address of a packet. That prefix determines the output port of the packet.

1.1.5   ERROR DETECTION

A node sends the bits of a packet to the next node by first converting them into electrical or optical signals. The receiving node converts the signals back into bits. This process is subject to errors caused by random fluctuations in the signals. Thus, it occasionally happens that some bits in a packet get corrupted, which corresponds to a *transmission error*.

A simple scheme to detect errors is for the source to add one bit, called *parity bit*, to the packet so that the number of ones is even. For instance, if the packet is 00100101, the sending node adds a parity bit equal to 1 so that the packet becomes 001001011 and has an even number of ones. If the receiver gets a packet with an odd number of ones, say 001101011, it knows that a transmission error occurred. This simple parity bit scheme cannot detect if two or any even number of bits were modified during the transmission. This is why the Internet uses a more robust error detection code for the packet headers, referred to as the *header checksum*. When using the header checksum, the sending node calculates the checksum bits (typically 16 bits) from the other fields in the header. The receiving node performs the same calculation and compares the result with the error detection bits in the packet; if they differ, the receiving node knows that some error occurred in the packet header, and it discards the corrupted packet.[3]

1.1.6   RETRANSMISSION OF ERRONEOUS PACKETS

In addition to dropping packets whose header is corrupted by transmission errors, a router along the path may discard arriving packets when it runs out of memory to temporarily store them before forwarding. This event occurs when packets momentarily arrive at a router faster than it can forward them to their next hop. Such packet losses are said to be due to *congestion*, as opposed to transmission errors.

To implement a reliable delivery, the source and destination use a mechanism that guarantees that the source retransmits the packets that do not reach the destination without errors. Such a scheme is called an *automatic retransmission request* or *ARQ*. The basic idea of this mechanism is that the destination acknowledges all the correct packets it receives, and the source retransmits packets that the destination did not acknowledge within a specific amount of time. Say that the source sends the packets 1, 2, 3, 4 and that the destination does not get packet 2. After a while, the source notices that the destination did not acknowledge packet 2 and retransmits a copy of that packet. We discuss the specific implementation of this mechanism in the Internet in the chapter on the transport protocol. Note that the source and destination hosts arrange for retransmissions, not the routers.

1.1.7   CONGESTION CONTROL

Imagine that many hosts send packets that happen to go though a common link in the network. If the hosts send packets too quickly, the link cannot handle them all, and the router with that outgoing link must discard some packets. To prevent an excessive number of discarded packets, the hosts slow down when they miss acknowledgments. That is, when a host has to retransmit a packet whose acknowledgment failed to arrive, it assumes that a congestion loss occurred and slows down the rate at which it sends packets.

Eventually, congestion subsides and losses stop. As long as the hosts get their acknowledgments in a timely manner, they slowly increase their packet transmission rate to converge to the maximum rate that can be supported by the prevailing network conditions. This scheme, called *congestion control*, automatically adjusts the transmission of packets so that the network links are fully utilized while limiting the congestion losses.

1.1.8   FLOW CONTROL

If a fast device sends packets very rapidly to a slow device, the latter may be overwhelmed. To prevent this phenomenon, the receiving device indicates, in each acknowledgment it sends back to the source, the amount of free buffer space it has to receive additional bits. The source stops

---

[3]We explain in the Transport chapter that the packet may also contain a checksum that the source calculates on the complete packet and that the destination checks to make sure that it does not ignore errors in the rest of the packet.

transmitting when this available space is not larger than the number of bits the source has already sent and the receiver has not yet acknowledged.

The source combines the *flow control* scheme with the congestion control scheme discussed earlier. Note that flow control prevents overflowing the destination buffer, whereas congestion control prevents overflowing router buffers.

1.2   DNS, HTTP, AND WWW

1.2.1   DNS

The hosts attached to the Internet have a *name* in addition to an IP address. The names are easier to remember (e.g., google.com). To send packets to a host, the source needs to know the IP address of that host. The Internet has an automatic directory service called the *Domain Name Service*, or *DNS*, that translates the name into an IP address. DNS is a distributed directory service. The Internet is decomposed into *zones*, and a separate *DNS server* maintains the addresses of the hosts in each zone. For instance, the department of EECS at Berkeley maintains the directory server for the hosts in the eecs.berkeley.edu zone of the network. The DNS server for that zone answers requests for the IP address of hosts in that zone. Consequently, if one adds a host on the network of our department, one needs to update only that DNS server.

1.2.2   HTTP AND WWW

The *World Wide Web* is arranged as a collection of hyperlinked resources such as web pages, video streams, and music files. The resources are identified by a *Uniform Resource Locator* or URL that specifies a computer and a file in that computer together with the protocol that should deliver the file.

For instance, the URL `http://www.eecs.berkeley.edu/~wlr.html` specifies a home page in a computer with name www.eecs.berkeley.edu and the protocol HTTP.

HTTP, the *Hyper Text Transfer Protocol*, specifies the request/response rules for getting the file from a server to a client. Essentially, the protocol sets up a connection between the server and the client, then requests the specific file, and finally closes the connection when the transfer is complete.

1.3   SUMMARY

- The Internet consists of hosts that send and/or receive information, routers, and links.

- Each host has a 32-bit IP address (in IPv4; 128-bit in IPv6) and a name. DNS is a distributed directory service that translates the name into an IP address.

- The hosts arrange the information into packets that are groups of bits with a specified format. A packet includes its source and destination addresses and error detection bits.

- The routers calculate the shortest paths (essentially) to destinations and store them in routing tables. The IP addresses are based on the location of the hosts to reduce the size of routing tables using longest prefix match.

- A source adjusts its transmissions to avoid overflowing the destination buffer (flow control) and the buffers of the routers (congestion control).

- Hosts remedy transmission and congestion losses by using acknowledgments, timeouts, and retransmissions.

1.4 PROBLEMS

P1.1 How many hosts can one have on the Internet if each one needs a distinct IPv4 address?

P1.2 If the addresses were allocated arbitrarily, how many entries should a routing table have?

P1.3 Imagine that all routers have 16 ports. In the best allocation of addresses, what is the size of the routing table required in each router?

P1.4 Assume that a host A in Berkeley sends a stream of packets to a host B in Boston. Assume also that all links operate at 100 Mbps and that it takes 130 ms for the first acknowledgment to come back after A sends the first packet. Say that A sends one packet of 1 KByte and then waits for an acknowledgment before sending the next packet, and so on. What is the long-term average bit rate of the connection? Assume now that A sends N packets before it waits for the first acknowledgment, and that A sends the next packet every time an acknowledgment is received. Express the long-term average bit rate of the connection as a function of N. [Note: 1 Mbps = 10^6 bits per second; 1 ms = 1 millisecond = 10^{-3} s.]

P1.5 Assume that a host A in Berkeley sends 1-KByte packets with a bit rate of 100 Mbps to a host B in Boston. However, B reads the bits only at 1 Mbps. Assume also that the device in Boston uses a buffer that can store 10 packets. Explore the flow control mechanism and provide a time line of the transmissions.

1.5 REFERENCES

Packet switching was independently invented in the early 1960s by Paul Baran [14], Donald Davies, and Leonard Kleinrock who observed, in his MIT thesis [56], that packet-switched networks can be analyzed using queuing theory. Bob Kahn and Vint Cerf invented the basic structure of TCP/IP in 1973 [52]. The congestion control of TCP was corrected by Van Jacobson in 1988 [49], partly motivated by the analysis by Chiu and Jain [25] and [26] and also by the stability of linear systems. Paul Mockapetris invented DNS in 1983. CIDR is described in [34]. Tim Berners-Lee invented the WWW in 1989. See [40] for a discussion of the Autonomous Systems.

CHAPTER 2

Principles

In networking, connectivity is the name of the game. The Internet connects a few billion computers across the world, plus associated devices such as printers, servers, and web cams. With the development of the Internet of Things, hundreds of billions of devices will soon be connected via the Internet. By "connect," we mean that the Internet enables these hosts to transfer files or bit streams among each other. To reduce cost, all these hosts share communication links. For instance, many users in the same neighborhood may share a coaxial cable to connect to the Internet; many communications share the same optical fibers. With this sharing, the network cost per user is relatively small.

To accommodate its rapid growth, the Internet is organized in a hierarchical way and adding hosts to it only requires local modifications. Moreover, the network is independent of the applications that it supports. That is, only the hosts need the application software. For instance, the Internet supports video conferences even though its protocols were designed before video conferences existed. In addition, the Internet is compatible with new technologies such as wireless communications or improvements in optical links.

We comment on these important features in this chapter. In addition, we introduce some metrics that quantify the performance characteristics of a network.

2.1 SHARING

Imagine designing a network to interconnect a few billion devices such as computers, smart phones, thermostats, etc. It is obviously not feasible to connect each pair of devices with a dedicated link. Instead, one attaches devices that are close to each other with a local network. One then attaches local networks together with an access network. One attaches the access networks together in a regional network. Finally, one connects the regional networks together via one or more backbone networks that go across the country, as shown in Figure 2.1.

This arrangement reduces considerably the total length of wires needed to attach the devices together, when compared with pairwise links. Many devices share links and routers. For instance, all the devices in the leftmost access network of the figure share the link L as they send information to devices in another access network. What makes this sharing possible is the fact that the devices do not transmit all the time. Accordingly, only a small fraction of devices are active and share the network links and routers, which reduces the investment required per device.

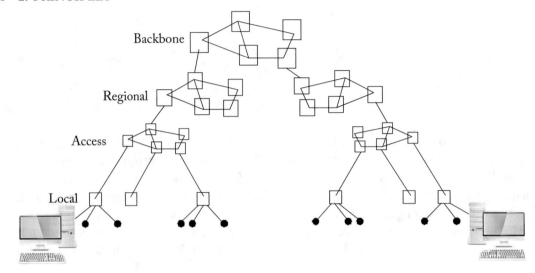

Figure 2.1: **A** hierarchy of networks.

As an example, you use your smart phone only a small fraction, say 1/100, of the time to access information from the Web. Accordingly, if a thousand users like you share a link of the network, only about ten are active at any given time. Hence, if the data rate of that link is C, then each user sees a data rate $C/10$. If all the users were active at the same time, each would get a data rate $C/1000$. The factor 100 is called the *multiplexing gain*. It measures the benefit of sharing a link among many users.

2.2   METRICS

To clarify the discussion of network characteristics, it is necessary to define precisely some basic metrics of performance.

2.2.1   LINK RATE

A link is characterized by a *rate*. For instance, a cable modem connection is characterized by two rates: one uplink rate (from the user's device to the Internet) and one downlink rate (from the Internet to the user's device). Typical values are 343 Mbps for the downlink rate and 131 Mbps for the uplink rate.

If the rate of a link is 100 Mbps, then the transmitter can send a packet with 10,000 bits in 0.1 ms. If it can send packets back-to-back without any gap, then the link can send a 100 MByte file in approximately 8 s. In practice, the protocols introduce gaps between packets.

A link that connects a user to the Internet is said to be *broadband* if its rate exceeds 25 Mbps downlink and 4 Mbps uplink. (This is the value defined by the Federal Communication Commission in 2015.) Otherwise, one says that the link is *narrowband*.

2.2.2   LINK BANDWIDTH AND CAPACITY

A signal of the form $V(t) = A \sin(2\pi f_0 t)$ makes f_0 cycles per second. We say that its *frequency* is f_0 Hz. Here, Hz stands for *Hertz* and means one cycle per second. For instance, $V(t)$ may be the voltage at time t as measured across the two terminals of a telephone line. The physics of a transmission line limits the set of frequencies of signals that it transports. The *bandwidth* of a communication link measures the width of that range of frequencies. For instance, if a telephone line can transmit signals over a range of frequencies from 300 Hz to 1 MHz ($= 10^6$ Hz), we say that its bandwidth is about 1 MHz.

The rate of a link is related to its bandwidth. Intuitively, if a link has a large bandwidth, it can carry voltages that change very quickly and it should be able to carry many bits per seconds, as different bits are represented by different voltage values. The maximum rate of a link depends also on the amount of noise on the link. If the link is very noisy, the transmitter should send bits more slowly. This is similar to having to articulate more clearly when talking in a noisy room.

An elegant formula due to Claude Shannon indicates the relationship between the *maximum reliable* link rate C, also called the *Shannon Capacity* of the link, its bandwidth W, and the noise. That formula is

$$C = W \log_2(1 + SNR).$$

In this expression, *SNR*, the *signal-to-noise ratio*, is the ratio of the power of the signal at the receiver divided by the power of the noise, also at the receiver. For instance, if $SNR = 10^6 \approx 2^{20}$ and $W = 1$ MHz, then we find that $C = 10^6 \log_2(1 + 10^6)$bps $\approx 10^6 \log_2(2^{20})$bps $= 10^6 \times 20$ bps $= 20$ Mbps. This value is the theoretical limit that could be achieved, using the best possible technique for representing bits by signals and the best possible method to avoid or correct errors. An actual link never quite achieves the theoretical limit, but it may come close.

The formula confirms the intuitive fact that if a link is longer, then its capacity is smaller. Indeed, the power of the signal at the receiver decreases with the length of the link (for a fixed transmitter power). For instance, a *Digital Subscriber Loop (DSL)* link over a telephone line that is very long has a smaller rate than a shorter line. The formula also explains why a coaxial cable can have larger rate than a telephone line if one knows that the bandwidth of the former is wider.

In a more subtle way, the formula shows that if the transmitter has a given power, it should allocate more power to the frequencies within its spectrum that are less noisy. Over a cello, one hears a soprano better than a basso profondo. A DSL transmitter divides its power into small frequency bands, and it allocates more power to the less noisy portions of the spectrum.

2.2.3   DELAY

Delay refers to the elapsed time for a packet to traverse between two points of interest. If the two points of interest are the two end points (i.e., a pair of communicating hosts), we refer to the delay between these two points as the end-to-end delay. End-to-end delay typically comprises the transmission and propagation times over the intervening links, and queuing and processing times at the intervening nodes (e.g., routers and switches). Figure 2.2 illustrates components of the overall delay for a packet of P bits traversing from node A to node B. In the scenario illustrated in the figure, when this packet is available to be sent over the link, there are Q other bits already present at node A that need to be sent over the link first. The link between node A and node B has the link rate of R bits per second (bps) and propagation time of T s. Queuing time refers to the waiting time of a packet at a node before it can be transmitted; transmission time refers to the time it takes for a packet to be transmitted over a link at the link rate; propagation time refers to the time for a physical signal to propagate over a link. In the scenario illustrated in the figure, queueing time and transmission time are Q/R and P/R seconds, respectively. One other kind of delay incurred by packets as they traverse the nodes in a network is processing time (not included in the illustration in Figure 2.2). Processing time refers to the time consumed in performing the required operations on a packet at a node. The concepts of queuing time and transmission time are further discussed in Sections 2.2.6 and 2.2.7.

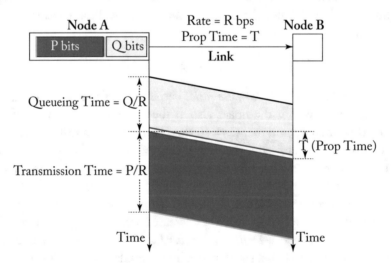

Figure 2.2: Queueing time, transmission time, and propagation time.

2.2.4   THROUGHPUT

Say that you download a large file from a server using the Internet. Assume that the file size is *B* bits and that it takes *T* s to download the file. In that case, we say that the *throughput* of the transfer is B/T bps. For instance, you may download an MP4 video file of 3 GBytes = 24 Gbits in 4 min, i.e., 240 s. The throughput is then 24 Gbits/240 s $= 0.1$ Gbps or 100 Mbps. (For convenience, we approximate 1 GByte by 10^9 bytes even though it is actually $2^{30} = 1,073,741,824$ bytes. Recall also that one defines 1 Mbps $= 10^6$ bps and 1 Kbps $= 10^3$ bps.)

The link rate is not the same as the throughput. For instance, in the example above, your video download had a throughput of 100 Mbps. You might have been using a cable modem with a downlink rate of 343 Mbps. The difference comes from the fact that the download corresponds to a sequence of packets and there are gaps between the packets.

Figure 2.3 shows two typical situations where the throughput is less than the link rate. The left part of the figure shows a source S that sends packets to a destination D. A packet is said to be *outstanding* if the sender has transmitted it but has not received its acknowledgment. Assume that the sender can have up to three outstanding packets. The figure shows that the sender sends three packets, then waits for the acknowledgment of the first one before it can send the fourth packet, and so on. In the figure, the transmitter has to wait because the time to get an acknowledgment—RTT, for *round-trip time*—is longer than the time the transmitter takes to send three packets. Because of that waiting time during which the transmitter does not send packets, the throughput of the connection is less than the rate R of the link. Note that, in this example, increasing the allowed number of outstanding packets increases the throughput until it becomes equal to the link rate. The maximum allowed number of outstanding bytes is called the *window size*. Thus, the throughput is limited by the window size divided by the round-trip time.

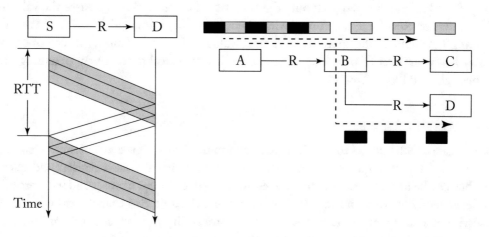

Figure 2.3: The throughput can be limited by the window size (left) and by a bottleneck link (right).

The right part of the figure shows devices A, C, D attached to a router B. Device A sends packets at rate R. Half the packets go to C and the other half go to D. The throughput of the connection from A to C is $R/2$ where R is the rate of the links. Thus, the two connections (from A to C and from A to D) share the rate R of the link from A to B. This link is the *bottleneck* of the system: it is the rate of that link that limits the throughput of the connections. Increasing the rate of that link would enable to increase the throughput of the connections.

2.2.5 DELAY JITTER

The successive packets that a source sends to a destination do not face exactly the same delay across the network. One packet may reach the destination in 50 ms whereas another one may take 120 ms. These fluctuations in delay are due to the variable amount of congestion in the routers. A packet may arrive at a router that has already many other packets to transmit. Another packet may be lucky and have rarely to wait behind other packets. One defines the *delay jitter* of a connection as the difference between the longest and shortest delivery time among the packets of that connection. For instance, if the delivery times of the packets of a connection range from 50 ms to 120 ms, the delay jitter of that connection is 70 ms.

Many applications such as streaming audio or video and voice-over-IP are sensitive to delays. Those applications generate a sequence of packets that the network delivers to the destination for playback. If the delay jitter of the connection is J, the destination should store the first packet that arrives for at least J seconds before playing it back, so that the destination never runs out of packets to play back. In practice, the value of the delay jitter is not known in advance. Typically, a streaming application stores the packets for T s, say $T = 4$, before starting to play them back. If the playback buffer gets empty, the application increases the value of T, and it buffers packets for T s before playing them back. The initial value of T and the rule for increasing T depend on the application. A small value of T is important for interactive applications such as voice over IP or video conferences; it is less critical for one-way streaming such as Internet radio, IPTV, or YouTube.

2.2.6 M/M/1 QUEUE

To appreciate effects such as congestion, jitter, and multiplexing, it is convenient to recall a basic result about delays in a queue. Imagine that customers arrive at a cash register and queue up until they can be served. Assume that one customer arrives in the next second with probability λ, independently of when previous customers arrived. Assume also that the cashier completes the service of a customer in the next second with probability μ, independently of how long he has been serving that customer and of how long it took to serve the previous customers. That is, λ customers arrive per second, on average, and the cashier serves μ customers per second, on average, when there are customers to be served. Note that the average service time per customer is $1/\mu$ s since the cashier can serve μ customers per second, on average. One defines the *utilization* of the system as the ratio $\rho = \lambda/\mu$. Thus, the utilization is 80% if the arrival rate is equal to 80% of the service rate.

Such a system is called an M/M/1 queue. In this notation, the first M indicates that the arrival process is memoryless: the next arrival occurs with probability λ in the next second, no matter how long it has been since the previous arrival. The second M indicates that the service is also memoryless. The 1 in the notation indicates that there is only one server (the cashier).

Assume that $\lambda < \mu$ so that the cashier can keep up with the arriving customers. The basic result is that the average time T that a customer spends waiting in line or being served is given by

$$T = \frac{1}{\mu - \lambda}.$$

If λ is very small, then the queue is almost always empty when a customer arrives. In that case, the average time T is $1/\mu$ and is equal to the average time the cashier spends serving a customer. Consequently, for any given $\lambda < \mu$, the difference $T - 1/\mu$ is the average *queuing time* that a customer waits behind other customers before getting to the cashier.

Another useful result is that the average number L of customers in line or with the server is given by

$$L = \frac{\lambda}{\mu - \lambda}.$$

Note that T and L grow without bound as λ increases and approaches μ.

To apply these results to communication systems, one considers that the customers are packets, the cashier is a transmitter, and the waiting line is a buffer attached to the transmitter. The average packet service time $1/\mu$ (in seconds) is the average packet length (in bits) divided by the link rate (in bps). Equivalently, the service rate μ is the link rate divided by the average length of a packet. Consider a computer that generates λ packets per second and that these packets arrive at a link that can send μ packets per second. The formulas above provide the average delay T per packet and the average number L of packets stored in the link's transmitter queue.

For concreteness, say that the line rate is 10 Mbps, that the packets are 1 Kbyte long, on average, and that 1,000 packets arrive at the transmitter per second, also on average. In this case, one finds that

$$\mu = \frac{10^7}{8 \times 10^3} = 1{,}250 \text{ packets per second.}$$

Consequently,

$$T = \frac{1}{\mu - \lambda} = \frac{1}{1{,}250 - 1{,}000} = 4 \text{ ms} \quad \text{and} \quad L = \frac{\lambda}{\mu - \lambda} = \frac{1{,}000}{1{,}250 - 1{,}000} = 4.$$

Thus, an average packet transmission time is $1/\mu = 0.8$ ms, the average delay through the system is 4 ms, and four packets are in the buffer, on average. A typical packet that arrives at the buffer finds 4 other packets already there; that packet waits for the 4 transmission times of those packets then for its own transmission time before it is out of the system. The average delay corresponds to 5 transmission times of 0.8 ms each, so that $T = 4$ ms. This delay T consists of the queuing time 3.2 ms and one transmission time 0.8 ms.

Because of the randomness of the arrivals of packets and of the variability of the transmission times, not all packets experience the same delay. With the M/M/1 model, it is, in principle, possible for a packet to arrive at the buffer when it contains a very large number of packets. However, that is not very likely. One can show that about 5% of the packets experience a delay larger than 12 ms and about 5% will face a delay less than 0.2 ms. Thus, most of the packets have a delay between 0.2 ms and 12 ms. One can then consider that the delay jitter is approximately 12 ms, or three times the average delay through the buffer.

To appreciate the effect of congestion, assume now that the packets arrive at the rate of 1,150 packets per second. One then finds that $T = 10$ ms and $L = 11.5$. In this situation, the transmission time is still 0.8 ms but the average queuing time is equal to 11.5 transmission times, or 9.2 ms. The delay jitter for this system is now about 30 ms. Thus, the average packet delay and the delay jitter increase quickly as the arrival rate of packets λ approaches the transmission rate μ.

Now imagine that N computers share a link that can send $N\mu$ packets per second. In this case, replacing μ by $N\mu$ and λ by $N\lambda$ in the previous formulas, we find that the average delay is now T/N and that the average number of packets is still L. Thus, by sharing a faster link, the packets face a smaller delay. This effect, which is not too surprising if we notice that the transmission time of each packet is now much smaller, is another benefit of having computers share fast links through switches instead of having slower dedicated links.

Summarizing, the M/M/1 model enables us to estimate the delay and backlog at a transmission line. The average transmission time (in seconds) is the average packet length (in bits) divided by the link rate (in bps). For a moderate utilization ρ (say $\rho \leq 80\%$), the average delay is a small multiple of the average transmission time (say 3–5 times). Also, the delay jitter can be estimated as about 3 times the average delay through the buffer.

2.2.7   LITTLE'S RESULT

Another basic result helps us understand some important aspects of networks: Little's Result, discovered by John D. C. Little in 1961. Imagine a system where packets arrive with an average rate of λ packets per second. Designate by L the average number of packets in the system and by W the average time that a packet spends in the system. Then, under very weak assumptions on the arrivals and processing of the packets, the following relation holds:

$$L = \lambda W.$$

Note that this relation, called *Little's Result*, holds for the M/M/1 queue, as the formulas indicate. However, this result holds much more generally.

To understand Little's Result, one can argue as follows. Imagine that a packet pays one dollar per second it spends in the system. Accordingly, the average amount that each packet pays is W dollars. Since packets go through the system at rate λ, the system gets paid at the average rate of λW per second. Now, this average rate must be equal to the average number L of packets in the system since they each pay at the rate of one dollar per second. Hence $L = \lambda W$.

As an illustration, say that 1 billion users send 10 MBytes per day each via the Internet, on average. Assume also that each packet spends 0.4 s in the Internet, on average. The average number of bits in transit on the Internet is then L where

$$L = \lambda W = \left(10^9 \times \frac{8 \times 10^7}{24 \times 3,600} \text{ bps}\right) \times 0.4\,\text{s} \approx 3.7 \times 10^{11} \text{ bits} \approx 43\,\text{Gbytes}.$$

Some of these bits are in transit in the fibers, the others are mostly in the routers.

As another example, let us try to estimate how many bits are stored in a fiber. Consider a fiber that transmits bits at the rate of 2.4 Gbps and the fiber is used at 20% of its capacity. Assume that the fiber is 100 km long. The propagation time of light in a fiber is about 5 μs per km. This is equal to the propagation time in a vacuum multiplied by the refractive index of the fiber. Thus, each bit spends a time W in the fiber where $W = 5\mu s \times 100 = 5 \times 10^{-4}$ s. The average arrival rate of bits in the fiber is $\lambda = 0.2 \times 2.4$ Gbps $= 0.48 \times 10^9$ bps. By Little's Result, the average number of bits stored in the fiber is L where

$$L = \lambda W = \left(0.48 \times 10^9 \text{ bps}\right) \times \left(5 \times 10^{-4} \text{ s}\right) = 2.4 \times 10^5 \text{ bits} = 30\,\text{Kbytes}.$$

A router also stores bits. To estimate how many, consider a router with 16 input ports at 1 Gbps. Assume the router is used at 10% of its capacity. Assume also each bit spends 5 ms in the router. We find that the average arrival rate of bits is $\lambda = 0.1 \times 16 \times 10^9$ bps. The average delay is $W = 5 \times 10^{-3}$ s. Consequently, the average number L of bits in the router is

$$L = \lambda W = \left(1.6 \times 10^9 \text{ bps}\right) \times \left(5 \times 10^{-3} \text{ s}\right) = 8 \times 10^6 \text{ bits} = 1\,\text{MByte}.$$

2.2.8   FAIRNESS

Consider two flows that share a link with rate 1 Mbps and assume that flow 1 values a given rate twice as much as flow 2. How should they share the link capacity? One possibility is to maximize the total value that the two flows derive from using the link. In this case, flow 1 should get the full capacity of the link, i.e., 1 Mbps, because it values it more than flow 2. This is called the *max-sum* allocation, or maximizing the *social welfare*. This solution does not seem fair to flow 2. Another possibility is to maximize the minimum *utility* of the two flows. If flow 1 gets a rate x and flow 2 a rate $1 - x$, the utility of flow 1 is $A2x$ for some constant A and that of flow 2 is $A(1 - x)$. To maximize the minimum, one chooses x so that $A2x = A(1 - x)$, i.e., $x = 1/3$ and both flows get a utility equal to $2A/3$. This is called the *max-min* allocation. Yet another possibility is to maximize the product of the utilities $2Ax$ and $A(1 - x)$. In this case, $x = 1/2$ and the utility of flow 2 is equal to A whereas that of flow 2 is $A/2$. This allocation is called *proportional fairness*. Achieving fairness across a network is more challenging than the simple example of the single link considered here. This can often lead to a trade-off between the network efficiency and fairness. We will explore this issue further in the chapter on models.

2.3   SCALABILITY

For a network to be able to grow to a large size, like the Internet, it is necessary that modifications have only a limited effect.

In the early days of the Internet, each computer had a complete list of all the computers attached to the network. Thus, adding one computer required updating the list that each computer maintained. Each modification had a global effect. Let us assume that half of the 1 billion computers on the Internet were added during the last ten years. That means that during these last ten years approximately $0.5 \times 10^9/(10 \times 365) > 10^5$ computers were added each day, on average.

Imagine that the routers in Figure 2.1 have to store the list of all the computers that are reached to each of their ports. In that case, adding a computer to the network requires modifying a list in all the routers, clearly not a viable approach.

Consider also a scheme where the network must keep track of which computers are currently exchanging information. As the network gets larger, the potential number of simultaneous communications also grows. Such a scheme would require increasingly complex routers.

Needless to say, it would be impossible to update all these lists to keep up with the changes. Another system had to be devised. We describe some methods that the Internet uses for scalability.

2.3.1   LOCATION-BASED ADDRESSING

We explained in the previous chapter that the IP addresses are based on the location of the devices, in a scheme similar to the telephone numbers.

One first benefit of this approach is that it reduces the size of the routing tables, as we already discussed. Figure 2.4 shows $M = N^2$ devices that are arranged in N groups with N devices each. Each group is attached to one router.

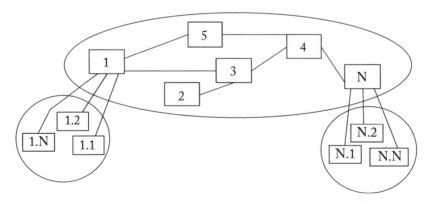

Figure 2.4: A simple illustration of location-based addressing.

In this arrangement, each router $1, \ldots, N$ needs one routing entry for each device in its group plus one routing entry for each of the other $N - 1$ routers. For instance, consider a packet that goes from device 1.2 to device N.1. First, the packet goes from device 1.2 to router 1. Router 1 has a routing entry that specifies the next hop toward router N, say router 5. Similarly, router 5 has a routing entry for the next hop, router 4, and so on. When the packet gets to router N, the latter consults its routing table to find the next hop toward device N.1. Thus, each routing table has size $N + (N - 1) \approx 2\sqrt{M}$.

In contrast, we will explain that the Ethernet network uses addresses that have no relation to the location of the computers. Accordingly, the Ethernet switches need to maintain a list of the computers attached to each of their ports. Such an approach works fine for a relatively small number of computers, say a few thousands, but does not scale well beyond that. If the network has M devices, each routing table needs M entries.

Assume that one day most electric devices will be attached to the Internet, such as light bulbs, coffee machines, door locks, curtain motors, cars, etc. What is a suitable way to select the addresses for these devices?

2.3.2   TWO-LEVEL ROUTING

The Internet uses a routing strategy that scales at the cost of a loss of optimality. The mechanism, whose details we discuss in the chapter on routing, groups nodes into domains.

Each domain uses a *shortest-path algorithm* where the length of a path is defined as the sum of the metrics of the links along the path. A faster link has a smaller metric than a slower link.

To calculate the shortest path between two nodes in a domain, the switches must exchange quite a bit of metric information. For instance, each node could make up a message with the list of its neighbors and the metrics of the corresponding links. The nodes can then send these messages to each other. For a simple estimate of the necessary number of messages, say that there are N routers and that each router sends one message to each of the other $N - 1$ routers. Thus, each router originates approximately N messages. For simplicity, say that each message goes through some average number of hops, say 6, that does not depend on N. The total number of messages that the N routers transmit is then about $6N^2$, or about $6N$ messages per router. This number of messages becomes impractical when N is very large.

To reduce the complexity of the routing algorithms, the Internet groups the routers into domains. It then essentially computes a shortest path across domains and each domain computes shortest paths inside itself. Say that there are N domains with N routers inside each domain. Each router participates in a shortest path algorithm for its domain, thus sending about $6N$ messages. In addition, one representative router in each domain participates in a routing algorithm across domains, also sending of the order of $6N$ messages. If the routing were done in one level, each router would transmit $6N^2$ messages, on average, which is N times larger than when using two-level routing.

In the chapter on routing, we explain another motivation for a two-level routing approach. It is rooted in the fact that inter-routing may correspond to economic arrangements as the domains belong to different organizations.

2.3.3   BEST EFFORT SERVICE

A crucial design choice of the Internet is that it does not guarantee any precise property of the packet delivery such as delay or throughput. Instead, the designers opted for simplicity and decided that the Internet should provide a "best effort service," which means that the network should try to deliver the packets as well as possible, whatever this means. This design philosophy is in complete contrast with that of the telephone network where precise bounds are specified for the time to get a dial tone, to ring a called party, and for the delay of transmission of the voice signal. For instance, voice conversation becomes very unpleasant if the one-way delay exceeds 250 ms. Consequently, how can one hope to use the Internet for voice conversations if it cannot guarantee that the delay is less than 250 ms? Similarly, a good video requires a throughput of at least 50 Kbps. How can one use the Internet for such applications? The approach of the Internet is that, as its technology improves, it will become able to support more and more demanding applications. Applications adapt to the changing quality of the Internet services instead of the other way around.

2.3.4   END-TO-END PRINCIPLE AND STATELESS ROUTERS

Best effort service makes it unnecessary for routers (or the network) to keep track of the number or types of connections that are set up. Moreover, errors are corrected by the source and destination that arrange for retransmissions. More generally, the guiding principle is that tasks should not be performed by routers if they can be performed by the end devices. This is called the *end-to-end principle*.

Thus, the routers perform their tasks for individual packets. The router detects if a given packet is corrupted and discards it in that case. The router looks at the packet's destination address to determine the output port. The router does not keep a copy of a packet for retransmission in the event of a transmission or congestion loss on the next hop. The router does not keep track of the connections, of their average bit rate or of any other requirement.

Accordingly, routers can be *stateless*: they consider one packet at a time and do not have any information about connections. This feature simplifies the design of routers, improves robustness, and makes the Internet scalable.

2.3.5   HIERARCHICAL NAMING

The Internet uses an automated directory service called DNS (for Domain Name System). This service translates the *name* of a computer into its address. For instance, when you want to connect to the Google server, the service tells your web browser the address that corresponds to the name www.google.com of the server.

The names of the servers, such as www.google.com, are arranged in a hierarchy as a tree. For instance, the tree splits up from the root into first-level domains such as .com, .edu, .be, and so on. Each first-level domain then splits up into second-level domains, etc. The resulting tree is then partitioned into subtrees or zones. Each zone is maintained by some independent administrative entity, and it corresponds to some directory server that stores the addresses of the computers with the corresponding names. For instance, one zone is eecs.berkeley.edu and another is stat.berkeley.edu and ieor.berkeley.edu.

The point of this hierarchical organization of DNS is that a modification of a zone only affects the directory server of that zone and not the others. That is, the hierarchy enables a distributed administration of the directory system in addition to a corresponding decomposition of the database.

2.4   APPLICATION AND TECHNOLOGY INDEPENDENCE

The telephone network was designed to carry telephone calls. Although it could also be used for a few other applications such as teletype and fax, the network design was not flexible enough to support a wide range of services.

The Internet is quite different from the telephone network. The initial objective was to enable research groups to exchange files. The network was designed to be able transport packets with a specified format including agreed-upon source and destination addresses. A source fragments a large file into chunks that fit into individual packets. Similarly, a bit stream can be transported by a sequence of packets. Engineers quickly designed applications that used the packet transport capability of the Internet. Examples of such applications include email, the World Wide Web, streaming video, voice over IP, peer-to-peer, video conferences, and social networks. Fundamentally, information can be encoded into bits. If a network can transport packets, it can transport any type of information.

2.4.1   LAYERS

To formalize this technology and application independence, the Internet is arranged into five functional *layers* as illustrated in Figure 2.5. The *Physical Layer* is responsible to deliver bits across a physical medium. The *Link Layer* uses the bit delivery mechanism of the Physical Layer to deliver packets across the link. The *Network Layer* uses the packet delivery across successive links to deliver the packets from source to destination. The *Transport Layer* uses the end-to-end packet delivery to transport individual packets or a byte stream from a process in the source computer to a process in the destination computer. The transport layer implements error control (through acknowledgments and retransmissions), congestion control (by adjusting the rate of transmissions to avoid congesting routers), and flow control (to avoid overflowing the destination buffer). Finally, the *Application Layer* implements applications that use the packet or byte stream transport service. A packet from the application layer may be a message or a file.

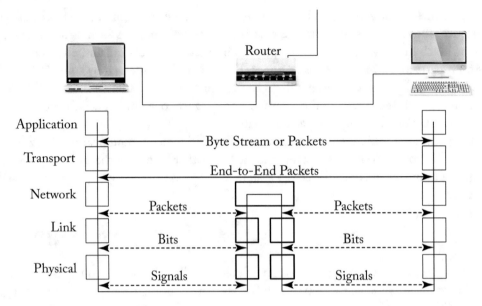

Figure 2.5: The five layers of Internet.

This layered decomposition provides compatibility of different implementations. For instance, one can replace a wired implementation of the Physical and Link Layers by a wireless implementation without having to change the higher layers and while preserving the connectivity with the rest of the Internet. Similarly, one can develop a new video conferencing application without changing the other layers. Moreover, the layered decomposition simplifies the design process by decomposing it into independent modules. For instance, the designer of the physical layer does not need to consider the applications that will use it. We explain in the chapter Models that the benefits of layering come at a cost of loss of performance.

2.5 APPLICATION TOPOLOGY

Networked applications differ in how hosts exchange information. This section explains the main possibilities. It is possible for a particular networked application to fit the description of more than one mode of information exchange described below.

2.5.1 CLIENT/SERVER

Web browsing uses a *client/server* model. In this application, the user host is the client and it connects to a server such as google.com. The client asks the server for files and the server transmits them. Thus, the connection is between two hosts and most of the transfer is from the server to the client.

In web browsing, the transfer is initiated when the user clicks on a hyperlink that specifies the server name and the files. In some cases, the request contains instructions for the server, such as keywords in a search or operations to perform, such as converting units or doing a calculation.

Popular websites are hosted by a *server farm*, which is a collection of servers (from hundreds to thousands) equipped with a system for balancing the load of requests. That is, the requests arrive at a device that keeps track of the servers that can handle them and of how busy they are; that device then dispatches the requests to the servers most likely to serve them fast. Such a device is sometimes called a *layer 7 router*. One important challenge is to adjust the number of active servers based on the load to minimize the energy usage of the server farm while maintaining the quality of its service.

2.5.2 P2P

A P2P (Peer-to-Peer) system stores files in user hosts instead of specialized servers. For instance, using BitTorrent, a user looking for a file (say a music MP3 file) finds a list of user hosts that have that file and are ready to provide it. The user can then request the file from those hosts. A number of hosts can deliver different fractions of that file in parallel.

The P2P structure has one major advantage over the client/server model: a popular file is likely to be available from many user hosts. Consequently, the service capacity of the system increases automatically with the popularity of files. In addition, the parallel download overcomes the asymmetry between upload and download link rates. Indeed, the Internet connection of a residential user is typically 3 times faster for downloads than for uploads. This asymmetry is justified for client/server applications. However, it creates a bottleneck when the server belongs to a residential user. The parallel downloads remove that asymmetry bottleneck.

2.5.3 CLOUD COMPUTING

Cloud computing refers to the new paradigm where a user makes use of the computing service hosted by a collection of computers attached to the network. A number of corporations make such services available. Instead of having to purchase and install the applications on his server, a user can lease the services from a cloud computing provider. The user can also upload his software on the cloud servers. The business driver for cloud computing is the sharing of the computing resources among different users. Provision of sufficient computing resources to serve a burst of requests satisfactorily is one of the key issues.

The servers in the cloud are equipped with software to run the applications in a distributed way. Some applications are well suited for such a decomposition, such as indexing of web pages and searches. Other applications are more challenging to distribute, such as simulations or extensive computations.

2.5.4 CONTENT DISTRIBUTION

A *content distribution system* is a set of servers (or server farms) located at various points in the network to improve the delivery of information to users. When a user requests a file from one server, that server may redirect the request to a server closer to the user. One possibility is for the server to return a homepage with links selected based on the location of the requesting IP address. Akamai is a content distribution system that many companies use.

2.5.5 MULTICAST/ANYCAST

Multicast refers to the delivery of files or streams to a set of hosts. The hosts subscribe to a multicast and the server sends them the information. The network may implement multicast as a set of one-to-one (unicast) deliveries. Alternatively, the network may use special devices that replicate packets to limit the number of duplicate packets that traverse any given link. Twitter is a multicast application.

Anycast refers to the delivery of a file to any one of a set of hosts. For instance, a request for information might be sent to any one of a set of servers.

2.5.6 PUSH/PULL

When a user browses the web, his host *pulls* information from a server. In contrast, *push* applications send information to user hosts. For instance, a user may subscribe to a daily newspaper that a server forwards according to a schedule when it finds the user's host available.

2.5.7 DISCOVERY

In most applications, the user specifies the files he is looking for. However, some applications discover and suggest information for the user. For instance, one application searches for users that are in the neighborhood and are looking for someone to have dinner with. Another application lists menus of nearby restaurants.

2.6 SUMMARY

The Internet has the following important features:

- Packet switching enables efficient sharing of network resources (statistical multiplexing);

- Hierarchical naming, location-based addressing, two-level routing, and stateless routers (possible because of the best effort service and the end-to-end principle) make the Internet scalable;

- Layering simplifies the design, and provides independence with respect to applications and compatibility with different technologies; and

- Applications have different information structures that range from client server to cloud computing to P2P to discovery.

  Some key metrics of performance of the Internet are as follows:

- The bandwidth (or bit rate) of a link;

- The throughput of a connection;

- The average delay and delay jitter of a connection; and

- Little's Result relates the backlog, the average delay, and the throughput.

2.7   PROBLEMS

P2.1 Assume 400 users share a 100 Mbps link. Each user is active a fraction 10% of the time, during a busy hour. If all the active users get an equal fraction of the link rate, what is the average throughput per active user?

P2.2 In this problem, we refine the estimates of the previous problem. If one flips a coin $n \gg 1$ times and each coin flip has a probability $p \in (0, 1)$ of yielding "Head," the average number of Heads is np. Moreover, the probability that the fraction of Heads is larger than $p + 1.3\sqrt{p(1 - p)/n}$ is about 10%. Use this fact to calculate the rate R such that the probability that the throughput per active user in the previous problem is less than R is only 10%.

P2.3 Imagine a switch in a network where packets arrive at the rate $\lambda = 10^6$ packets per second. Assume also that a packet spends $T = 1 \text{ ms} = 10^{-3}$ s in the switch, on average. Calculate the average number of packets that the switch holds at any given time.

P2.4 Packets with an average length of 1 KBytes arrive at a link to be transmitted. The arrival rate of packets corresponds to 8 Mbps. The link rate is 10 Mbps. Using the M/M/1 delay formula, estimate the average delay per packet. What fraction of that delay is due to queuing?

P2.5 Say that a network has N domains with M routers each and K devices are attached to each router. Assume that the addresses are not geographically based but are assigned randomly to the devices instead. How many entries are there in the routing table of each router if the routing uses only one level? What if the network uses a 2-level routing scheme? Now assume that the addresses are based on location. What is the minimum average size of the routing table in each router?

P2.6 Consider a router in the backbone of the Internet. Assume that the router has 24 ports, each attached to a 1 Gbps link. Say that each link is used at 15% of its capacity by connections that have an average throughput of 200 Kbps. How many connections go through the router at any given time? Say that the connections last an average of 1 min. How many new connections are set up that go through the router in any given second, on average?

P2.7   We would like to transfer 20 Kbyte file across a network from node A to node F. Packets have a length of 1 Kbyte (neglecting header). The path from node A to node F passes through 5 links, and 4 intermediate nodes. Each of the links is a 10 km optical fiber with a rate of 10 Mbps. The 4 intermediate nodes are store-and-forward devices, and each intermediate node must perform a 50 μs routing table look up after receiving a packet before it can begin sending it on the outgoing link. How long does it take to send the entire file across the network?

P2.8   Suppose we would like to transfer a file of K bits from node A to node C. The path from node A to node C passes through two links and one intermediate node, B, which is a store-and-forward device. The two links are of rate R bps. The packets contain P bits of data and a 6 byte header. What value of P minimizes to time it takes to transfer the file from A to C? Now suppose there are N intermediate nodes, what value of P minimizes the transfer time in this case?

P2.9   Consider the case of GSM cell phones. GSM operates at 270.88 Kbps and uses a spectrum spanning 200 KHz. What is the theoretical SNR (in dB) that these phones need for operation? In reality, the phones use an SNR of 10 dB. Use Shannon's theorem to calculate the theoretical capacity of the channel, under this signal-to-noise ratio. How does the utilized capacity compare with the theoretical capacity?

P2.10   Packet switching uses resources efficiently by taking advantage of the fact that only a fraction of potential senders are active at any time. In this problem, you will demonstrate this fact mathematically. Suppose we have a single link with capacity L bits/second and a population of users that generate data at r bits/second when busy. The probability that a user is busy generating data is p.

   (a)   What is the maximum number of users that can be supported using circuit switching? Call this value M_c.

   (b)   Now suppose we use packet switching to support a population of M_p users. Derive a formula (in terms of p, M_p, N, L, and r) for the probability that more than N users are busy.

   (c)   Plug in some numbers. Let $L = 1$ Mbps, $r = 64$ Kbps, and $p = 0.1$. Find the value for M_c. What is the probability that more than M_c users are busy for $M_p = 2 * M_c$? What about $M_p = 4 * M_c$?

P2.11   Consider two links in series as shown in the Figure 2.6. Link 1 has the transmission rate of R_1 bps and the propagation delay of T_1 s, while link 2 has the transmission rate of R_2 bps and the propagation delay of T_2 s. Assume $R_1 < R_2$. Consider a packet of size (in bits)

Figure 2.6: Figure for principles Problem 11.

$L = H + D$, where H is the size of the header (in bits) and D is the size of the data (in bits).

(a) Suppose the Switch does store-and-forward for the packets transmitted from the Source to the Destination. Find how long it takes between the time the first bit of the packet begins transmission at the Source and the last bit of the packet reaches the Destination. Assume no processing delays at the Switch.

(b) Suppose the Switch does cut-through switching. In cut-through switching, the Switch waits for the complete header to arrive before transmitting the packet onto the next link. Assume there are no processing delays between receiving the header and transmitting the packet on link 2.

   i. Since $R_2 > R_1$, it is possible that, during cut-through switching, link 2 would need to wait for the next bit of the packet to arrive over link 1, and thus causing gaps. Find the maximum amount of data D that can be sent in the packet without facing this problem.

   ii. Now assume that the size of data D is below the threshold calculated in part (b)i above. How long does it take between the time the first bit of the packet begins transmission at the Source and the last bit of the packet reaches the Destination?

P2.12 Suppose that the source has a file of size S bits to send to the destination. Assume that all transmission between the source and destination follows a sequence of L links, with $L - 1$ intervening switches as shown in the Figure 2.7. Each of the switches on the path does store and forward and each of the links transmits at a rate of R bps. Assume that the propagation delay on the links is negligible. Also, assume that the packets do not suffer any switching or queuing delay at the switches.

Figure 2.7: Figure for principles Problem 12.

(a) Suppose the source breaks the file into N chunks of size $D = S/N$ bits, and attaches a header of size H before transmitting each of these chunks as a packet. Thus, each packet is of size $P = (D + H)$ bits. Calculate the time T it takes between the start of transmission of the first packet at the source and the time at which the last packet is completely received at the destination.

(b) Now, use the expression for T obtained above to determine the optimal D at which T is minimized.

(c) What happens to optimal D obtained above when

    i. $H \approx 0$? Why?

    ii. $H \gg S$? Why?

2.8   REFERENCES

The theory of channel capacity is due to Claude Shannon [92]. A fun tutorial on queuing theory can be found at [94]. Little's Result is proved in [60]. For a discussion of two-level routing, see [34]. The end-to-end principle was elaborated in [88]. The structure of DNS is explained in [72]. The layered structure of the Internet protocols is discussed in [21].

CHAPTER 3

Ethernet

Ethernet is a technology used to connect up to a few hundred computers and devices. The connections use wires of up to 100 m or fibers of up to a few km. The bit rate on these wires and fibers is usually 100 Mbps but can go to 1 Gbps or even 10 Gbps. The vast majority of computers on the Internet are attached to an Ethernet network or, increasingly, its wireless counterpart WiFi. We discuss WiFi in a separate chapter.

We first review a typical Ethernet network. We then explore the history of Ethernet. We then explain the addressing and frame structure. After a brief discussion of the physical layer, we examine switched Ethernet. We then discuss Aloha and hub-based Ethernet.

3.1   TYPICAL INSTALLATION

Figure 3.1 shows a typical installation. The computers are attached to a *hub* or to a *switch* with wires or fibers. We discuss hubs and switches later in this chapter. We call *ports* the different attachments of a switch or a hub (same terminology as for routers). A small switch or hub has 4 ports; a large switch has 48 or more ports. The hub is used less commonly now because of the superiority of the switch. The figure also shows a wireless extension of the network.

3.2   HISTORY OF ETHERNET

Ethernet started as a wired version of a wireless network, the Aloha network. Interestingly, after two decades, a wireless version of Ethernet, WiFi was developed.

3.2.1   ALOHA NETWORK

The Aloha network was invented around 1970 by Norman Abramson and his collaborators. He considered the scenario where multiple devices distributed over different Hawaiian islands wanted to communicate with the mainframe computer at the main campus in Honolulu. Figure 3.2 shows an Aloha network. The devices transmit packets to a central communication node attached to the mainframe computer on frequency f_1. This node acknowledges packets on frequency f_2. A device concludes that its packet is not delivered at the destination if it does not receive an acknowledgment within a prescribed time interval. The procedure for transmission is as follows: If a device is not waiting for an acknowledgment of a previously transmitted packet and has a packet to transmit, it transmits it immediately. If the device does not get an acknowledgment within a prescribed time interval, it retransmits the packet after a random time interval.

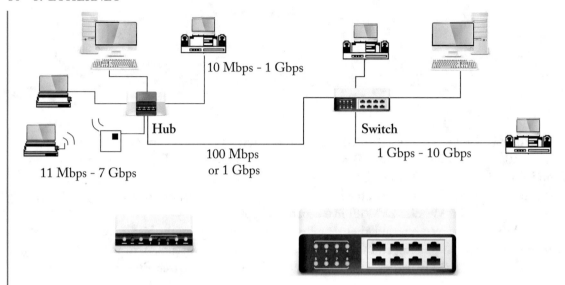

Figure 3.1: Typical Ethernet installation.

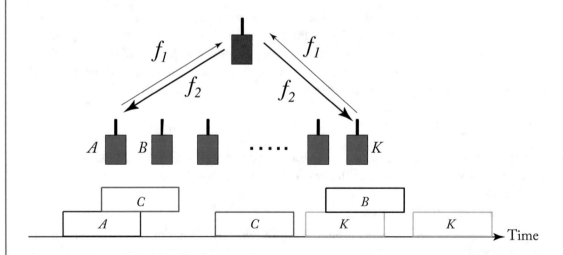

Figure 3.2: Aloha network and the timing of transmissions.

The system uses two frequencies so that a device can receive acknowledgments regardless of any transmissions. Since the receiver is overwhelmed by the powerful signal that the device transmits, it cannot tell if another device is also transmitting at the same time. Consequently, a device transmits a complete packet even though this packet may *collide* with another transmission, as shown in the timing diagram.

The major innovation of the Aloha network is *randomized multiple access*. Using this mechanism, the device transmissions are not scheduled ahead of time. Instead, the devices resolve their conflicts with a distributed and randomized mechanism.

The mechanism of the Aloha network is as follows:

- If an acknowledgment is not outstanding, transmit immediately;

- If no acknowledgment, repeat after a random delay.

3.2.2 CABLE ETHERNET

Between 1973 and 1976, Robert Metcalfe and his colleagues developed a wired version of the Aloha network illustrated in Figure 3.3. This version became known as *Ethernet*. All the devices

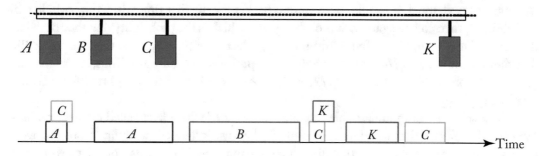

Figure 3.3: A cable Ethernet.

share a common cable. They use a randomized multiple access protocol similar to the Aloha network but with two key differences. A device waits for the channel to be idle before transmitting, and listens while it transmits and it aborts its transmission as soon as it detects a collision. These *carrier sensing* and *collision detection* mechanisms reduce the time wasted by collisions.

Thus, the mechanism of this network is as follows:

- (Wait) For a new packet, waiting time is set to 0. For a retransmission, waiting time is set to a random time according to the backoff scheme described below.

- (Transmit) After the waiting time, when the channel is idle, transmit.

- (Collision Detection) If a collision is detected, abort transmission and go to (Wait) for a retransmission. Otherwise, got to (Wait) for a new packet.

The stations choose their random waiting time (also called *backoff time*) as a multiple X of one time slot. A *time slot* is defined to be 512 bit transmission times for 10 Mbps and 100 Mbps Ethernet and 4,096 bit times for Gbps Ethernet. A station picks X uniformly in $\{0, 1, \ldots, 2^n - 1\}$ where $n = \min\{m, 10\}$ and m is the number of collisions the station experienced with the same packet. When m reaches 16, the station gives up and declares an error. This scheme is called the *truncated binary exponential backoff*. Thus, the first time a station attempts to transmit, it does so just when the channel becomes idle (more precisely, it must wait a small specified interframe gap). After one collision, the station picks X to be equal to 0 or 1, with probability 0.5 for each possibility. It then waits for the channel to be idle and for $X \times 512$ bit times (at 10 Mbps or 100 Mbps). When the channel is sensed to be idle following this waiting time, the station transmits while listening, and it aborts when it detects a collision. A collision would occur if another station had selected the same value for X. It turns out that 512 bit times is longer than $2PROP$, so that if stations pick different values for X, the one with the smallest value transmits and the others hear its transmission before they attempt to transmit.

The main idea of this procedure is that the waiting time becomes more "random" after multiple collisions, which reduces the chances of repeated collisions. Indeed, if two stations are waiting for the channel to be idle to transmit a packet for the first time, then they collide with probability 1, then collide again with probability 1/2, which is the probability that they both pick the same value of X in $\{0, 1\}$. The third time, they select X in $\{0, 1, 2, 3\}$, so that the probability that they collide again is 1/4. The fourth time, the probability is only 1/8. Thus, the probability that the stations collide k times is $(1/2) \times (1/4) \times \cdots \times (1/2^{k-1})$ for $k \leq 11$, which becomes small very quickly.

This scheme has an undesirable (possibly unintended) side effect called *capture* or *winner takes all*. This effect is that a station that is unlucky because it happens to collide tends to have to keep on waiting as other stations that did not collide transmit. To see this, consider two stations A and B. Assume that they both have packets to transmit. During the first attempt, they collide. Say that A picks $X = 0$ and B picks $X = 1$. Then station A gets to transmit while B waits for one time slot. At the end of A's transmission, A is ready with a new packet and B is ready with its packet since it has decremented its backoff counter from the initial value 1 to 0 while it was waiting. Both stations then collide again. This was the first collision for the new packet of A, so that station picks X in $\{0, 1\}$. However, this was the second collision for B's packet, so that station picks X in $\{0, 1, 2, 3\}$. It is then likely that A will pick a smaller value and get to transmit again. This situation can repeat multiple times. This annoying problem never got fixed because Ethernet moved to a switched version.

3.2.3   HUB ETHERNET

For convenience of installation, Rock, Bennett, and Thaler developed the hub-based Ethernet (then called StarLAN) in the mid 1980s. In such a network shown in Figure 3.4 the devices are attached to a hub with a point-to-point cable. When a packet arrives at the hub, it repeats it on all its other ports. If two or more packets arrive at the hub at the same time, it sends a signal indicating that it has detected a collision on all its output ports. (The cables are bi-directional.) The devices use the same mechanism as in a cable-based Ethernet.

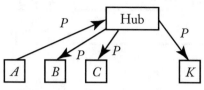 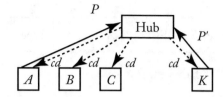

Figure 3.4: A hub-based Ethernet.

3.2.4 SWITCHED ETHERNET

In 1989, the company Kalpana introduced the *Ethernet switch*. As Figure 3.5 shows, the switch sends a packet only to the output port toward the destination of the packet. The main benefit of

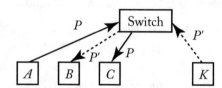

Figure 3.5: A switched Ethernet.

a switch is that multiple packets can go through the switch at the same time. Moreover, if two packets arrive at the switch at the same time and are destined to the same output port, the switch can store the packets until it can transmit them. Thus, the switch eliminates the collisions.

3.3 ADDRESSES

Every computer Ethernet attachment is identified by a globally unique 48-bit string called an *Ethernet Address* or *MAC Address* (for *Media Access Control*). This address is just an identifier since it does not specify where the device is. You can move from Berkeley to Boston with your laptop, and its Ethernet address does not change: it is hardwired into the computer.

Because the addresses are not location-based, the Ethernet switches maintain tables that list the addresses of devices that can be reached via each of its ports.

The address with 48 ones is the *broadcast address*. A device is supposed to listen to broadcast packets. In addition, devices can subscribe to *group multicast addresses*.

3.4 FRAME

Ethernet packets have the frame format shown in Figure 3.6. We show the *Ethernet II frame format*, the most commonly used Ethernet frame format in practice.

The frame format has the following fields:

7	1	6	6	2	46 - 1500	4 bytes
PRE	SFD	DA	SA	Type	Data	CRC

Figure 3.6: Ethernet frame.

- A 7-byte *preamble*. This preamble consists of alternating ones and zeroes. It is used for synchronizing the receiver.

- A 1-byte *start of frame delimiter*. This byte is 10101011, and it indicates that the next bit is the start of the packet.

- A 6-byte *source address*. This is source MAC address.

- A 6-byte *destination address*. This is destination MAC address.

- A 2-byte *Type* field. Type value in this field is always larger than or equal to 1,536, and it indicates the type of payload. For instance, Field 0x80 0x00 indicates TCP/IP; and that packet specifies its own length in a field inside the IP header. We note that in an alternate Ethernet frame format specified by the IEEE 802.3 standard, this field is used for specifying the length of the payload to allow for any type of payload. However, since the maximum payload length supported by Ethernet is 1,500 bytes, this field will always have a value smaller than or equal to 1,500 if the IEEE 802.3 frame format is used. This observation allows the Ethernet II and IEEE 802.3 frame formats to coexist.

- A 4-byte *cyclic redundancy check*. This is used for detecting frame corruption.

3.5   PHYSICAL LAYER

There are many versions of the physical layer of Ethernet. The name of a version has the form [rate][modulation][media or distance]. Here are the most common examples:

- 10Base5 (10 Mbps, baseband, coax, 500 m);

- 10Base-T (10 Mbps, baseband, twisted pair);

- 100Base-TX (100 Mbps, baseband, 2 pair);

- 100Base-FX (100 Mbps, baseband, fiber);

- 1000Base-CX for 2 pairs balanced copper cabling;

- 1000Base-LX for long wavelength optical transmission;

- 1000Base-SX for short wavelength optical transmission.

3.6   SWITCHED ETHERNET

3.6.1   EXAMPLE

Figure 3.7 shows a network with three switches and ten devices. Each device has a unique address. We denote these addresses by $A, B, \ldots, J$.

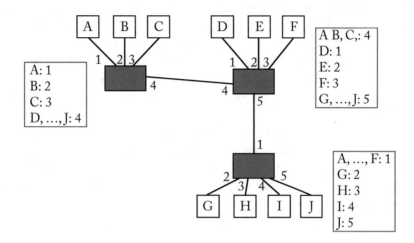

Figure 3.7: Example of switched Ethernet network.

3.6.2   LEARNING

When a computer with Ethernet address x sends a packet to a computer with Ethernet address y, it sends a packet $[y|x|data]$ to the switch. When it gets the packet, the switch reads the destination address y, looks in a table, called forwarding table, to find the port to which y is attached and sends the packet on these wires.

The switch updates its forwarding table as follows. When it gets a packet with a destination address that is not in the table, the switch sends a copy of the packet on all the ports, except the port on which the packet came. Whenever it gets a packet, it adds the corresponding source address to the table entry that corresponds to the port on which it came. Thus, if packet $[y|x|data]$ arrives to the switch via port 3, the switch adds an entry $[x = 3]$ to its table.

To select a unique path for packets in the network, Ethernet switches use the Spanning Tree Protocol, as explained below. Moreover, the network administrator can restrict the links on which some packets can travel by assigning the switch ports to virtual LANs. A virtual LAN, or VLAN, is an identifier for a subset of the links of an Ethernet network. When using VLANs, the Ethernet packet also contains a VLAN identifier for the packet. The source of the packet specifies that VLAN. Each switch port belongs to a set of VLANs that the network administrator configures. Packets are restricted to the links of its VLAN. The advantage of VLANs is that they separate an Ethernet network into distinct virtual networks, as if these networks did not share switches and links. This separation is useful for security reasons.

3.6.3   SPANNING TREE PROTOCOL

For reliability, it is common to arrange for a network to have multiple paths between two devices. However, with redundant paths, switches can loop traffic and create multiple copies.

To avoid these problems, the switches run the *spanning tree protocol* that selects a unique path between devices. Using this protocol, the switches find the tree of shortest paths rooted at the switch with the smallest identification number (ID). The ID of a switch is the smallest of the Ethernet addresses of its interfaces. What is important is that these IDs are different. To break up ties, a switch selects, among interfaces with the same distance to the root, the next hop with the smallest ID.

The protocol operates as follows. (See Figure 3.8.) The switches send packets with the information [*myID|CurrentRootID|DistancetoCurrentRootRoot*]. A switch relays packets whose Current Root ID is the smallest the switch has seen so far and the switch adds one to the distance to current root. Eventually, the switches only forward packets from the switch with the smallest ID with their distance to that switch.

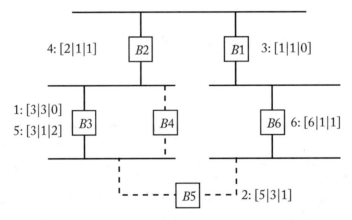

Figure 3.8: Example of messages when the spanning tree protocol runs.

The figure shows that, first, switch *B*3 sends packet [3|3|0] meaning "I am 3, I think the root is 3, and my distance to 3 is 0." Initially, the switches do not know the network. Thus, *B*3 does not know that there is another switch with a smaller ID. In step 2, *B*5 who has seen the first packet sends [5|3|1] because it thinks that the root is 3 and it is 1 step away from 3. The different packet transmissions are not synchronized. In the figure, we assume that happens in that order. Eventually, *B*1 sends packet [1|1|0] that the other switches forward.

To see how ties are broken, consider what happens when *B*3 and *B*4 eventually relay the packet from *B*1. At step 5, the figure shows that *B*3 sends [3|1|2] on its bottom port. At some later step, not shown in the figure, $B4$ sends [4|1|2]. Switch $B5$ must choose between these two paths to connect to $B1$: either via $B3$ or via $B4$ that are both two steps away from $B1$. The tie is broken in favor of the switch with the smaller ID, thus $B3$ in this example. The result of the spanning tree protocol is shown in Figure 3.8 by indicating the active links using solid lines.

Summing up, the spanning tree protocol avoids loops by selecting a unique path in the network. Some switches that were installed for redundancy get disabled by the protocol. If an active switch or link fails, the spanning tree protocol automatically selects a new tree.

Note that, although the spanning tree is composed of shortest paths to the root, the resulting routing may be far from optimal. To see this, imagine a network arranged as a ring with $2N + 1$ switches. The two neighboring switches that are at distance N from the switch with the smallest ID communicate via the longest path.

3.7   ALOHA

In this section, we explore the characteristics of the *Aloha network*. We start with a time-slotted version of the protocol and then we study a non-slotted version.

3.7.1   TIME-SLOTTED VERSION

Consider the following *time-slotted* version of the Aloha protocol. Time is divided into slots. The duration of one time slot is enough to send one packet. Assume there are n stations and that the stations transmit independently with probability p in each time slot. The probability that exactly one station transmits in a given time slot is (see Section 3.10)

$$R(p) := np(1 - p)^{n-1}.$$

The value of p that maximizes this probability can be found by setting the derivative of $R(p)$ with respect to p equal to zero. That is,

$$0 = \frac{d}{dp} R(p) = n(1 - p)^{n-1} - n(n - 1)p(1 - p)^{n-2}.$$

Solving this equation, we find that the best value of p is

$$p = \frac{1}{n}.$$

This result confirms the intuition that p should decrease as n increases. Replacing p by $1/n$ in the expression for $R(p)$, we find that

$$R\left(\frac{1}{n}\right) = \left(1 - \frac{1}{n}\right)^{n-1} \approx \frac{1}{e} \approx 0.37. \tag{3.1}$$

The last approximation comes from the fact that

$$\left(1 + \frac{a}{n}\right)^n \approx e^a \text{ for } n \gg 1.$$

The quality of the approximation (3.1) is shown in Figure 3.9.

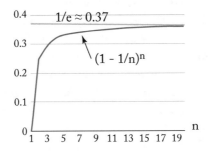

Figure 3.9: The approximation for (3.1).

Thus, if the stations are able to adjust their probability p of transmission optimally, the protocol is successful at most 37% of the time. That is, 63% of the time slots are either idle or wasted by transmissions that collide.

3.8   NON-SLOTTED ALOHA

So far, we assumed that all the Aloha stations were synchronized. What if they are not and can instead transmit at arbitrary times? This version of Aloha is called *non-slotted* or *pure*. The interesting result is that, in that case, they can only use a fraction $1/2e \approx 18\%$ of the channel. To model this situation, consider very small time slots of duration $\tau \ll 1$. One packet transmission time is still equal to one unit of time. Say that the stations start transmitting independently with probability p in every small time slot and then keep on transmitting for one unit of time. The situation was shown in Figure 3.2.

The analysis in the appendix shows that by optimizing over p one gets a success rate of at most 18%.

3.9   HUB ETHERNET

The study of hub-based Ethernet is somewhat involved. The first step is to appreciate the maximum time T it takes to detect a collision. This results in a particular randomization procedure, which incurs a wasted waiting time equal to an integer multiple of T. From that understanding, we can calculate the efficiency.

3.9.1   MAXIMUM COLLISION DETECTION TIME

Imagine that two nodes A and B try to transmit at about the same time. Say that A start transmitting at time 0. (See Figure 3.10.) The signal from A travels through the wires to the hub which repeats it. The signal then keeps on travelling toward B. Let *PROP* indicate the maximum propagation time between two devices in this network. By time *PROP*, the signal from

A reaches *B*. Now, imagine that *B* started transmitting just before time *PROP*. It thought the system was idle and could then transmit. The signal from *B* will reach *A* after a time equal to *PROP*, that is just before time 2*PROP*. A little bit later, node *A* will detect a collision. Node *B* detected the collision around time *PROP*, just after it started transmitting. To give a chance to *A* to detect the collision, *B* does not stop as soon as it detects the collision. This might result in *B* sending such a short signal that *A* might ignore it. Instead, *B* sends a "jam" signal, long enough to have the energy required for *A* to notice it.

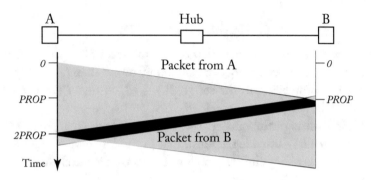

Figure 3.10: The maximum time for *A* to detect a collision is 2*PROP*.

The nodes wait a random multiple of $T = 2PROP$ before they start transmitting and they transmit if the system is still idle at the end of their waiting time. The point is that if they choose different multiples of T after the system last became idle, then they will not collide. To analyze the efficiency of this system, assume that n stations transmit independently with probability p in each time slot of duration T. We know that the probability that exactly one station transmits during one time slot is at most $1/e$. Thus, as shown in the appendix, the average time until the first success is e time slots of duration T. After this success, one station transmits successfully for some average duration equal to *TRANS*, defined as the average transmission time of a packet.

Thus, every transmission with duration *TRANS* requires a wasted waiting time of $(e - 1) \times T = 2(e - 1) \times PROP$. The fraction of time during which stations transmit successfully is then

$$\eta = \frac{TRANS}{2(e - 1)PROP + TRANS} \approx \frac{1}{1 + 3.4A}$$

where $A = PROP/TRANS$. We note that extensive Ethernet simulations show that $\eta \approx 1/(1 + 5A)$. The difference may be because the backoff scheme used in practice does not achieve the ideal randomization assumed in the analysis.

3.10   APPENDIX: PROBABILITY

In our discussion of the Aloha protocol, we needed a few results from Probability Theory. This appendix provides the required background.

3.10.1 PROBABILITY

Imagine an experiment that has N equally likely outcomes. For instance, one rolls a balanced die and the six faces are equally likely to be selected by the roll. Say that an event A occurs when the selected outcome is one of M of these N equally likely outcomes. We then say that the probability of the event A is M/N and we write $P(A) = M/N$.

For instance, in the roll of the die, if the event A is that one of the faces $\{2, 3, 4\}$ is selected, then $P(A) = 3/6$.

3.10.2 ADDITIVITY FOR EXCLUSIVE EVENTS

For this example, say that the event B is that one of the faces $\{1, 6\}$ is selected. Note that the events A and B are *exclusive*: they cannot occur simultaneously. Then "A or B" is a new event that corresponds to the outcome being in the set $\{1, 2, 3, 4, 6\}$, so that $P(A \text{ or } B) = 5/6$. Observe that $P(A \text{ or } B) = P(A) + P(B)$.

In general, it is straightforward to see that if A and B are exclusive events, then $P(A \text{ or } B) = P(A) + P(B)$. This idea is illustrated in the left part of Figure 3.11. Moreover, the same property extends to any finite number of events that are exclusive two by two. Thus, if $A_1, \ldots, A_n$ are exclusive two by two, then

$$P(A_1 \text{ or } A_2 \cdots \text{ or } A_n) = P(A_1) + P(A_2) + \cdots + P(A_n).$$

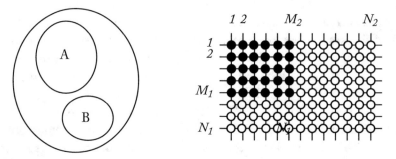

Figure 3.11: Additivity for exclusive events (left) and product for independent events (right).

Indeed, if event A_m corresponds to M_m outcomes for $m = 1, \ldots, n$ and if the sets of outcomes of these different events have no common element, then the event A_1 or $A_2 \cdots$ or A_n corresponds to $M_1 + \cdots + M_n$ outcomes and

$$\begin{aligned}
P(A_1 \text{ or } A_2 \cdots \text{ or } A_n) &= \frac{M_1 + \cdots + M_n}{N} = \frac{M_1}{N} + \cdots + \frac{M_n}{M} \\
&= P(A_1) + \cdots + P(A_n).
\end{aligned}$$

3.10.3 INDEPENDENT EVENTS

Now consider two experiments. The first one has N_1 equally likely outcomes and the second has N_2 equally likely outcomes. The event A is that the first experiment has an outcome that is in a set of M_1 of the N_1 outcomes. The event B is that the second experiment has an outcome in a set of M_2 of the N_2 outcomes. Assume also that the two experiments are performed "independently" so that the $N_1 \times N_2$ pairs of outcomes of the two experiments are all equally likely. Then we find that the event "A and B" corresponds to $M_1 \times M_2$ possible outcomes out of $N_1 \times N_2$, so that

$$P(A \text{ and } B) = \frac{M_1 \times M_2}{N_1 \times N_2} = \frac{M_1}{N_1} \times \frac{M_2}{N_2} = P(A)P(B).$$

Thus, we find that if A and B are "independent," then the probability that they both occur is the product of their probabilities. (See right part of Figure 3.11.)

For instance, say that we roll two balanced dice. The probability that the first one yields an outcome in $\{2, 3, 4\}$ and that the second yields an outcome in $\{3, 6\}$ is $(3/6) \times (2/6)$.

One can generalize this property to any finite number of such independent experiments. For instance, say that one rolls the die three times. The probability that the three outcomes are in $\{1, 3\}$, $\{2, 4, 5\}$, and $\{1, 3, 5\}$, respectively, is $(2/6) \times (3/6) \times (3/6)$.

3.10.4 SLOTTED ALOHA

Recall the setup of slotted Aloha. There are n stations that transmit independently with probability p in each time slot. We claim that the probability of the event A that exactly one station transmits is

$$P(A) = np(1 - p)^{n-1}. \tag{3.2}$$

To see this, for $m \in \{1, 2, \ldots, n\}$, define A_m to be the event that station m transmits and the others do not. Note that

$$A = A_1 \text{ or } A_2 \ldots \text{ or } A_n$$

and the events $A_1, \ldots, A_n$ are exclusive. Hence,

$$P(A) = P(A_1) + \cdots + P(A_n). \tag{3.3}$$

Now, we claim that

$$P(A_m) = p(1 - p)^{n-1}, m = 1, \ldots, n. \tag{3.4}$$

Indeed, the stations transmit independently. The probability that station m transmits and the other $n - 1$ stations do not transmit is the product of the probabilities of those events, i.e., the product of p and $1 - p$ and $\ldots$ and $1 - p$, which is $p(1 - p)^{n-1}$.

Combining the expressions (3.4) and (3.3) we find (3.2).

3.10.5 NON-SLOTTED ALOHA

Recall that in pure Aloha, we consider that the stations start transmitting independently with probability p in small time slots of duration $\tau \ll 1$. A packet transmission lasts one unit of time, i.e., $1/\tau$ time slots. With τ very small, this model captures the idea that stations can start transmitting at any time.

Consider one station, say station 1, that transmits a packet P, as shown in Figure 3.12. The set of starts of transmissions that collide with P consists of $2/\tau - 1$ time slots: the $1/\tau - 1$ time slots that precede the start of packet P and the $1/\tau$ time slots during the transmission of P. Indeed, any station that would start transmitting during any one of these $2/\tau - 1$ time slots

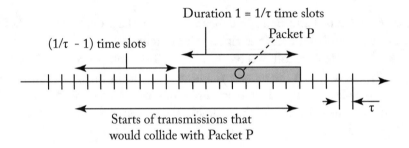

Figure 3.12: The starts of transmissions that collide with packet P cover $2/\tau - 1$ time slots.

would collide with P. Accordingly, the probability $S(p)$ that station 1 starts transmitting in a given small time slot and is successful is the probability it starts transmitting and that no other station starts transmitting in $2/\tau - 1$ small time slots. That probability is then

$$S(p) = p(1 - p)^{(n-1)(2/\tau-1)}.$$

In this expression, the first factor p is the probability that station 1 starts transmitting in a particular time slot, and the other $(n - 1)(2/\tau - 1)$ factors $(1 - p)$ are the probability that the other $n - 1$ stations do not transmit in each of the $2/\tau - 1$ time slots that would conflict with P.

The average number of transmissions per unit of time is then $nS(p)/\tau$ since there are $1/\tau$ slots per unit of time and n stations that might succeed. The value of p that maximizes $S(p)$ is such that the derivative $S'(p)$ of $S(p)$ with respect to p is equal to zero. That is, with $2/\tau - 1 = K$,

$$0 = \frac{d}{dp}S(p) = (1 - p)^{(n-1)K} - (n - 1)Kp(1 - p)^{(n-1)K-1}.$$

This gives

$$p^* = [(n - 1)K + 1]^{-1}.$$

The resulting success rate is

$$\frac{1}{\tau} n S(p^*) = \frac{K+1}{2} n S(p^*) = \frac{K+1}{2} n \frac{1}{(n-1)K+1} (1 - \frac{1}{(n-1)K+1})^{(n-1)K}.$$

With $M := (n-1)K \approx n(K+1) \approx (n-1)K + 1$, we find that the success rate is approximately equal to

$$\frac{1}{2} \left(1 - \frac{1}{M}\right)^M \approx \frac{1}{2e} \approx 18\%.$$

3.10.6 WAITING FOR SUCCESS

Consider the following experiment. You flip a coin repeatedly. In each flip, the probability of "Head" is p. How many times do you have to flip the coin, on average, until you get the first "Head"?

To answer this question, let us look at a long sequence of coin flips. Say that there are on average $N - 1$ "Tails" before the next "Head." Thus, there is one "Head" out of N coin flips, on average. The fraction of "Heads" is then $1/N$. But we know that this fraction is p. Hence, $p = 1/N$, or $N = 1/p$.

As another example, one has to roll a die six times, on average, until one gets a first outcome equal to "4." Similarly, if the probability of winning the California lottery is one in one million per trial, then one has to play one million times, on average, before the first win. Playing once a week, one can expect to wait about nineteen thousand years.

3.10.7 HUB ETHERNET

As we explained in the section on Hub Ethernet, the stations wait a random multiple of $T = 2PROP$ until they attempt to transmit. In a simplified model, one may consider that the probability that one station transmits in a given time slot with duration $2PROP$ is approximately $1/e$. Thus, the average time until one station transmits alone is e time slots. Of this average time, all but the last time slot is wasted. That is, the stations waste $(e - 1)$ time slots of duration $2PROP$ for every successful transmission, on average. Thus, for every transmission with duration $TRANS$, there is a wasted time $(e - 1) \times 2PROP \approx 3.4PROP$.

Consequently, the fraction of time when the stations are actually using the network to transmit packets successfully is $TRANS$ divided by $TRANS + 3.4PROP$. Thus, the efficiency of the Hub Ethernet is

$$\frac{TRANS}{TRANS + 3.4PROP}.$$

3.11 SUMMARY

Ethernet is a widely used networking technology because it is low cost and fast. The chapter reviews the main operations of this technology.

- Historically, the first multiple access network where nodes share a transmission medium was the Aloha network. A cable-based multiple access network followed, then a star-topopoly version that used a hub, then the switch replaced the hub.

- Each Ethernet device attachment has a unique 48-bit address.

- Multiple versions of the Ethernet physical layer exist, with rates that range from 10 Mbps to 10 Gbps.

- An Ethernet switch learns the list of devices attached to each port from the packet source addresses. Switches run a spanning tree protocol to avoid loops.

- The efficiency of slotted Aloha is at most 37% and that of non-slotted Aloha is at most 18%. That of a hub-based Ethernet decreases with the transmission rate and increases with the average packet length.

- The appendix reviews the basic properties of probabilities: the additivity for exclusive events and the product for independent events. It applies these properties to the analysis of the efficiency of simple MAC protocols.

3.12   PROBLEMS

P3.1 We have seen several calculations showing that the efficiency of an Ethernet random access scheme is well below 100%. Suppose we knew that there are exactly N nodes in the Ethernet. Here's a strategy: we divide the time into N slots and make the 1st node use the 1st slot, 2nd node use 2nd slot and so on (this is called time division multiplexing). This way, we could achieve 100% efficiency and there would never be any collision!! What's the problem with this plan?

P3.2 Consider a random access MAC protocol like Slotted ALOHA. There are N nodes sharing a media, and time is divided into slots. Each packet takes up a single slot. If a node has a packet to send, it always tries to send it out with a given probability. A transmission succeeds if a single node is trying to access the media and all other nodes are silent.

   (a) Suppose that we want to give differentiated services to these nodes. We want different nodes to get a different share of the media. The scheme we choose works as follows: If node i has a packet to send, it will try to send the packet with probability p_i. Assume that every node has a packet to send all the time. In such a situation, will the nodes indeed get a share of the media in the ratio of their probability of access?

   (b) Suppose there are 5 nodes, and the respective probabilities are $p_1 = 0.1$, $p_2 = 0.1$, $p_3 = 0.2$, $p_4 = 0.2$, and $p_5 = 0.3$. On an average, what are the probabilities that each node is able to transmit successfully, in a given time slot?

(c) Now suppose that nodes do not always have a packet to send. In fact, the fraction of time slots when a node has a packet to send (call it busy time b_i) is the same as its probability of access, i.e., $b_i = p_i$. For simplicity's sake, do not consider any queuing or storing of packets—only that node i has a packet to send on b_i of the slots. In such a situation, is the share of each node in the correct proportion of its access probability or busy time?

P3.3 In recent years, several networking companies have advocated the use of Ethernet (and VLANs) for networks far beyond a "local" area. Their view is that Ethernet as a technology could be used for much wider areas like a city (Metro), or even across several cities. Suggest two nice features of Ethernet that would still be applicable in a wider area. Also suggest two other characteristics which would not scale well, and would cause problems in such architectures.

P3.4 Consider a commercial 10 Mbps Ethernet configuration with one hub (i.e., all end stations are in a single collision domain).

(a) Find Ethernet efficiency for transporting 512 byte packets (including Ethernet overhead) assuming that the propagation delay between the communicating end stations is always 25.6 μs, and that there are many pairs of end stations trying to communicate.

(b) Recall that the maximum efficiency of Slotted Aloha is 1/e. Find the threshold for the frame size (including Ethernet overhead) such that Ethernet is more efficient than Slotted Aloha if the fixed frame size is larger than this threshold. Explain why Ethernet becomes less efficient as the frame size becomes smaller.

P3.5 Ethernet standards require a minimum frame size of 512 bits in order to ensure that a node can detect any possible collision while it is still transmitting. This corresponds to the number of bits that can be transmitted at 10 Mbps in one roundtrip time. It only takes one propagation delay, however, for the first bit of an Ethernet frame to traverse the entire length of the network, and during this time, 256 bits are transmitted. Why, then, is it necessary that the minimum frame size is 512 bits instead of 256?

P3.6 Consider the corporate Ethernet shown in Figure 3.13.

(a) Determine which links get deactivated after the Spanning Tree Algorithm runs, and indicate them on the diagram by putting a small X through the deactivated links.

(b) A disgruntled employee wishes to disrupt the network, so she plans on unplugging central Bridge 8. How does this affect the spanning tree and the paths that Ethernet frames follow?

P3.7 In the Figure 3.14, all of the devices want to transmit at an average rate of R Mbps, with equal amounts of traffic going to every other node. Assume that all of the links are half-duplex and operate at 100 Mbps and that the media access control protocol is perfectly

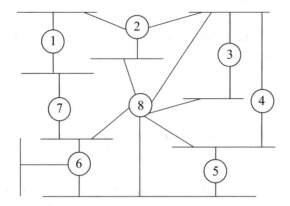

Figure 3.13: **Figure for Ethernet Problem 6.**

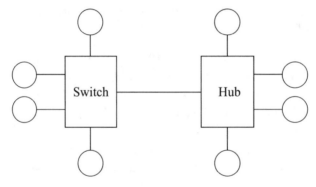

Figure 3.14: **Figure for Ethernet Problem 7.**

efficient. Thus, each link can only be used in one direction at a time, at 100 Mbps. There is no delay to switch from one direction to the other.

(a) What is the maximum value of R?

(b) The hub is now replaced with another switch. What is the maximum value of R now?

3.13   REFERENCES

The Aloha network is described and analyzed in [6]. Ethernet is introduced in [70]. One key innovation is the protocol where stations stop transmitting if they detect a collision. The actual mechanism for selecting a random time before transmitting (the so-called exponential backoff scheme) is another important innovation. For a clear introduction to probability, see [38].

CHAPTER 4

WiFi

The WiFi Alliance is the industrial consortium that decides implementations of the "wireless Ethernet" standards IEEE 802.11. The networks based on these standards are commonly referred to as the WiFi networks. In this chapter, we explain the basics of a WiFi network, including the multiple access scheme used by such a network.

4.1 BASIC OPERATIONS

In this chapter, we only describe the most widely used implementation of WiFi: the infrastructure mode with the *Distributed Coordination Function* (DCF). WiFi devices (laptops, printers, some cellular phones, etc.) are equipped with a radio that communicates with an *Access Point* (AP) connected to the Internet. The set of devices that communicate with a given access point is called a *Basic Service Set* (BSS). The AP advertises on a periodic basis (e.g., every 100ms) the identity, capabilities, and parameters of the associated BSS using a special broadcast frame called a beacon frame. The main components of WiFi are the MAC Sublayer and the Physical Layer.

The MAC Sublayer uses a binary exponential backoff. A station that gets a correct packet sends an acknowledgment (ACK) after waiting a short time interval. To send a packet a station must wait for a longer time interval plus the random backoff delay, so that it will never interfere with the transmission of an ACK. Finally, a variation of the protocol calls for a station that wants to send a packet P to first transmit a short *Request-to-Send* (RTS) to the destination that replies with a short *Clear-to-Send* (CTS). Both the RTS and CTS indicate the duration of the transmission of P and of the corresponding ACK. When they hear the RTS and the CTS, the other stations refrain from transmitting until P and its ACK have been sent. Without the RTS/CTS, some stations in the BSS might not hear the transmission of P and might end up interfering with it. We discuss these ideas further later in the chapter.

Originally, the Physical Layer was designed to operate in the *unlicensed band* (2.4 GHz or 5 GHz). More recent versions of WiFi also added use of sub-GHz and 60 GHz bands for specific applications. WiFi uses a sophisticated physical layer, including modulation schemes that support a transmission rate from 1 Mbps to multi-Gbps, depending on the version. In the last section of this chapter, we briefly describe the key aspects of some of the more recent versions of WiFi. Before that, we limit our discussion to the commonly used version of WiFi in practice. Each version uses a number of disjoint frequency channels, which enables the network manager to select different channels for neighboring BSSs to limit interference. The radios adjust their transmission rate to improve the throughput. When a few packets are not acknowledged, the sender reduces the transmission rate, and it increases the rate again after multiple successive packets get acknowledged.

4.2   MEDIUM ACCESS CONTROL (MAC)

We first describe the MAC protocol, then discuss the addresses that WiFi uses.

4.2.1   MAC PROTOCOL

The MAC protocol uses a few constants defined as follows in Table 4.1.

Table 4.1: Constants for the MAC protocol

Constant	802.11b	802.11a
Slot Time	20 μs	9 μs
SIFS	10 μs	16 μs
DIFS	50 μs	34 μs
EIFS	364 μs	94 μs
CW_{min}	31	15
CW_{max}	1023	1023

When a station (a device or the access point) receives a correct WiFi packet, it sends an ACK after an *SIFS* (for Short Inter-frame Spacing).

To send a packet, a station must wait for the channel to be idle for *DIFS* (for DCF Inter-Frame Spacing) plus the random backoff delay. Since *DIFS* > *SIFS*, a station can send an ACK without colliding with a packet. The backoff delay is z Slot Times, where z is picked uniformly in $\{0, 1, \ldots, CW_n\}$ where

$$CW_n = \min(CW_{\max}, (CW_{\min} + 1) * 2^n - 1), n = 0, 1, 2, \ldots \quad (4.1)$$

where n is the number of previous attempts at transmitting the packet. For instance, in 802.11b, $CW_0 = 31, CW_1 = 63, CW_2 = 127$, and so on. The value of z is used to initialize the backoff counter for the station. The station waits until the channel is idle for *DIFS*, and subsequently decrements the backoff counter by one every Slot Time during which the channel is idle. If another station transmits, it freezes its backoff counter, and resumes decrementing after the channel has been idle for *DIFS* again. The station transmits when the backoff counter reaches 0.

Each station attempts to decode each WiFi packet transmitted on the channel. It is only after decoding a packet that a station can determine if the received packet is intended for itself.

If a station receives an incorrect WiFi packet, it must wait for *EIFS* (for Extended Inter-Frame Spacing) before attempting to transmit in order to allow for the possibility that the intended receiver of that packet did receive it correctly. The value of *EIFS* is computed taking into account the possibility that the intended receiver might send an ACK at the lowest data rate allowed by the standard. These considerations lead to the values for *EIFS* shown in Table 4.1. The station receiving an incorrect WiFi packet resumes decrementing the residual value of its backoff counter after the channel has been idle for *EIFS*. If a correct packet is received during *EIFS*, the station reverts back to the normal operation of waiting for the channel to be idle for just *DIFS* before resumption of countdown of its backoff counter.

Figure 4.1 illustrates the key aspects of the WiFi MAC protocol. For this illustration, we use the parameters associated with the 802.11b version of the standard. There are two WiFi devices A and B, and one access point X. It is assumed that during the preceding successful transmission (i.e., successful transmission of a WiFi data packet and the corresponding ACK), the stations A, B, and X determine that each has a new packet to transmit on the channel. They wait for the channel to be idle for *DIFS*, and since they all have a new packet to transmit, they each generate a random number picked uniformly in $\{0, 1, \ldots, 31\}$ of Slot Times for their respective backoff delays. Since the backoff delay of B is the smallest in this illustration, station B transmits its packet, and after *SIFS*, the destination (the access point X) sends an ACK. The stations A and X freeze their backoff counters during the transmission of the packet by B and of the ACK by X. After the channel has remained idle for $DIFS$, the stations A and X resume decrementing their backoff counters, and B generates a new random backoff delay, picking uniformly a random number in $\{0, 1, \ldots, 31\}$. In this illustration, the backoff counters of A and B reach 0 at the same time. This causes A and B to start transmitting at the same time, resulting in a collision. The collision causes a corrupted reception at X, and that in turn causes A and B not to receive any ACK. We explain below how the WiFi MAC protocol continues after this collision.

We first note that each WiFi data packet or ACK transmission consists of the physical layer overheads and a WiFi MAC frame (see Figure 4.3). After transmitting a WiFi data packet, the transmitting station expects to start receiving a MAC ACK frame after *SIFS* plus the time for the physical layer overheads. For verifying that this actually happens, it sets an ACK Time-Out (TO) interval equal to the sum of *SIFS*, the time for the physical layer overheads, and a margin of a Slot Time (to account for processing and propagation times, etc.). For 802.11b, this leads to the value of 222 μs for ACK TO. As is the case in our illustration, if the start of any MAC frame does not occur during this interval, the transmitting station can conclude that it would need to retransmit the data packet it just attempted to transmit. For completeness, we note that if the start of a MAC frame does occur during ACK TO, the transmitting station waits for additional time to verify that the received MAC frame is indeed a valid ACK. In our scenario, A and B wait for the ACK TO duration after their respective transmissions, and then upon concluding the need for retransmission, they invoke the random backoff procedure by each

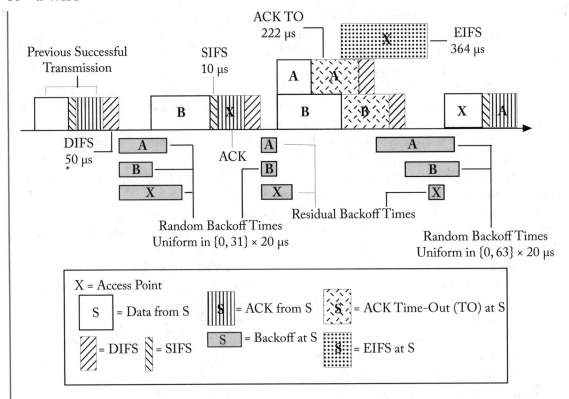

Figure 4.1: Key aspects of WiFi MAC (with 802.11b parameters).

generating a random number of Slot Times for its backoff delay by uniformly picking a random number in $\{0, 1, \ldots, 63\}$ (since each of them has already made one attempt for its current data packet). On the other hand, X waits for *EIFS* before resuming to decrement its residual backoff counter. In this illustration, the backoff counter of only X reaches 0 first. At that time, station X transmits (to A), and after *SIFS*, A sends an ACK.

4.2.2   ENHANCEMENTS FOR MEDIUM ACCESS

The MAC protocol described above is a *Carrier Sense Multiple Access with Collision Avoidance* (CSMA/CA) protocol. In general, wireless networks based on CSMA can suffer from two potential issues: (i) *Hidden terminal problem*, and (ii) *Exposed terminal problem*. The hidden terminal problem refers to the situation where two sending devices cannot hear each other (i.e., they are hidden from each other), and hence one of them can begin transmission even if the other device is already transmitting. This can result in a collision at the receiving device. In infrastructure mode WiFi networks, this issue can manifest itself in the situation where two devices hidden from each other end up having overlapping transmissions, and the AP failing to receive either of those transmissions. The exposed terminal problems refers to the situation where a device refrains from transmitting as it senses the medium to be busy, but its intended destination is different from that of the ongoing transmission, and hence could have in fact transmitted without any collision. In infrastructure mode WiFi networks, since all communication takes place via the AP, the reader can convince oneself that the exposed terminal problem is not really an issue.

To overcome the hidden terminal problem, WiFi networks can make use of the RTS and CTS messages. The idea is that a sender wanting to transmit data, first sends out an RTS message obeying the rules of medium access described above. This RTS message carries the address of the receiver to whom the sender wishes to send data. Upon receiving the RTS message, the intended receiver replies with a CTS message. Subsequently, upon receiving the CTS message, the sender begins data transmission which is followed by an ACK by the receiver. The entire sequence of RTS, CTS, data, and ACK messages are exchanged as consecutive messages separated by SIFS. Any device receiving either the RTS or CTS message deduces that the channel is going to be busy for the impending data transfer. Consider the situation where two devices hidden from each other want to send data to the AP. If the first device sends out an RTS message and the AP responds with a CTS message, the other device should hear at least the CTS message. Hence, the RTS/CTS exchange makes the devices hidden from the sending device aware of the busy channel status.

The MAC Sublayer also provides another indication of how long the channel is going to be busy. This is done by including the Duration field in the frame header (see Figure 4.2). This field is used to update *Network Allocation Vector* (NAV) variable at each device. The NAV variable indicates how long the channel is going to remain busy with the current exchange. Use of NAV for keeping track of channel status is referred to as *Virtual Carrier Sensing*.

4.2.3   MAC ADDRESSES

Figure 4.2, adapted from [42], shows the MAC frame format. We briefly describe below the fields in a frame. See [42] and [35] for further details.

Bytes:	2	2	6	6	6	2	6	0-2312	4
	Frame Control	Duration or ID	Address 1	Address 2	Address 3	Sequence Control	Address 4	Frame Payload	FCS

Figure 4.2: Generic MAC frame format.

The Frame Control field is used to indicate the type of frame (e.g., data frame, ACK, RTS, CTS, etc.) and some other frame parameters (e.g., whether more fragments of the higher layer frame are to follow, whether the frame is a retransmitted frame, etc.). The Duration field is used for setting NAV as mentioned above. Observe that the frame header contains up to four MAC addresses. Address 1 and Address 2 are the MAC addresses of the receiver and the transmitter, respectively. For extended LAN communication, Address 3 is the MAC address of source or destination device depending on whether the frame is being transmitted or received by the AP. Address 4 is used only in the extended LAN application where a WiFi network is used as a wireless bridge. In a typical home installation, the AP device also acts as a router. Here, only Address 1 and Address 2 are important as Address 3 coincides with one of these two addresses. It should be noted that the control frames (e.g., RTS, CTS, and ACK frames) and management frames (e.g., association request and beacon frames) do not need to use all the four addresses. The Sequence Control field is used to indicate the higher layer frame number and fragment number to help with defragmentation and detection duplicate frames. Finally, the FCS field provides the frame checksum to verify the integrity of a received frame.

4.3   PHYSICAL LAYER

WiFi networks operate in the unlicensed spectrum bands. In the U.S., these bands are centered around 2.4 or 5 GHz. Table 4.2 lists different versions of WiFi networks and their key attributes.

Table 4.2: Different types of WiFi networks

Key Standards	Max Rate	Spectrum (U.S.)	Year
802.11	2 Mbps	2.4 GHz	1997
802.11a	54 Mbps	5 GHz	1999
802.11b	11 Mbps	2.4 GHz	1999
802.11g	54 Mbps	2.4 GHz	2003
802.11n	600 Mbps	2.4 & 5 GHz	2009
802.11ad	7 Gbps	60 GHz	2012
802.11ac	6.8 Gbps	5 GHz	2013
802.11ah	234 Mbps	sub-GHz	2016

To get an idea of the frequency planning used, let us consider the IEEE 802.11b-based WiFi networks. Here the available spectrum around 2.4 GHz is divided into 11 channels with 5 MHz separation between the consecutive center frequencies. In order to minimize co-channel interference, channels 1, 6, and 11 are commonly used, giving channel separation of 25 MHz. Adjacent BSSs typically use different channels with maximum separation to minimize interference. Optimal channel assignment for different BSSs in a building or a campus is a non-trivial problem. In a relatively large building or campus (e.g., a department building on a university campus), it is common to have an *Extended Service Set* (ESS) created by linking BSSs. In such an environment, it is possible for a device to move from one BSS to the next in a seamless fashion.

WiFi networks are based on complex Physical Layer technologies. For example, the WiFi networks based on the IEEE 802.11a standard make use of *Orthogonal Frequency Division Multiplexing* (OFDM) in the 5GHz band. See [35, 43, 46, 46] for further details on the Physical Layer. The IEEE 802.11g standard introduced OFDM in the 2.4 GHz band for achieving a higher data rate. We will briefly discuss the more recent standards listed in Table 4.2 in the Recent Advances section at the end of this chapter.

4.4   EFFICIENCY ANALYSIS OF MAC PROTOCOL

In this section, we examine the *efficiency* of the WiFi MAC protocol, defined as the data throughput that the stations can achieve. We first consider the case of a single active station. The analysis determines the data rate and concludes that only 58% of the bit rate is used for data, the rest being overhead or idle time between frames. We then examine the case of multiple stations and explain a simplified model due to Bianchi [17].

4.4.1   SINGLE DEVICE

To illustrate the efficiency of WiFi networks, we first consider the scenario where a single device is continuously either transmitting to or receiving from the AP 1,500 bytes of data payload without using the RTS/CTS messages. Note that since only one device is involved, there are no channel collisions. We neglect propagation delay and assume that there are no losses due to channel errors. In this illustration, we consider an IEEE 802.11b-based WiFi network. Recall that for such a network, the basic system parameters are as follows: Channel Bit Rate = 11 Mbps, SIFS = $10\,\mu s$, Slot Time = $20\,\mu s$, and DIFS = $50\mu s$. As described in the medium access rules, after each successful transmission, the device performs a backoff for the number of Slot Times uniformly chosen in the set $\{0, 1, \ldots, 31\}$. Figure 4.3 shows how the channel is used in this scenario. Preamble and *Physical Layer Convergence Procedure* (PLCP) Header shown here are the Physical Layer overheads transmitted at 1 Mbps.

Let us calculate the data throughput for this scenario neglecting the propagation time. First, observe that the average backoff time for each frame corresponds to the average of 15.5 Slot Times, i.e., $15.5 * 20 = 310\,\mu s$. From Figure 4.3, note that the other overheads for each data frame are DIFS, Physical Layer overhead for data frame, MAC header and CRC, SIFS, Physical Layer overhead for ACK, and MAC Sublayer ACK. One can now easily calculate that the overheads mentioned above, including the backoff time, equal to a total of $788.9\,\mu s$ on average, and that the data payload takes $1500 * 8/11 = 1090.9\,\mu s$. This amounts to the data throughput of $1500 * 8/(788.9 + 1090.9) = 6.38$ Mbps, or overall efficiency of $6.38/11 = 58\%$.

4.4.2   MULTIPLE DEVICES

G. Bianchi [17] has provided an insightful analysis for the WiFi MAC protocol. He considers the scenario with n WiFi devices who always have data to transmit, and analyzes the Markov

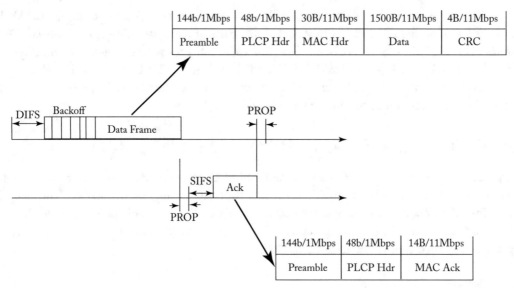

Figure 4.3: Channel usage for single user scenario.

chain $\{s(k), b(k)\}$ for a given device where $s(k)$ denotes the number of previous attempts for transmitting the pending packet at this device, and $b(k)$ denotes the backoff counter, both at the epoch k. (See the appendix for a brief tutorial on Markov chains.) This analysis also applies to the scenario where $n - 1$ devices and the AP have data to transmit. The Markov chain is embedded at the epoch k's when the device adjusts its backoff counter. The transition probabilities of this Markov chain are depicted in Figure 4.4, adapted from [17]. W_i and m here relate to $CW_{\min}$ and $CW_{\max}$ in (4.1) as follows: $W_i = 2^i(CW_{\min} + 1)$ and $CW_{\max} + 1 = 2^m(CW_{\min} + 1)$.

    To understand this diagram, consider a station that gets a new packet to transmit. The state of that station is the node at the top of the diagram. After waiting for *DIFS*, the station computes a backoff value uniformly in $\{0, 1, \ldots, 31\}$. The state of the station is then one of the states $(0, 0), (0, 1), \ldots, (0, 31)$ in the diagram. The first component of the state (0) indicates that the station has made no previous attempt at transmitting the pending packet. The second component of the state is the value of the backoff counter from the set $\{0, 1, \ldots, 31\}$. The station then decrements its backoff counter at suitable epochs (when no other station transmits and the channel is idle for a Slot Time, or after a *DIFS* and a Slot Time[1] following the channel busy condition). For instance, the state can then move from $(0, 5)$ to $(0, 4)$. Eventually, the backoff counter gets to 0, so that the state is $(0, 0)$ and the station transmits. If the transmission collides, then the station computes another backoff value X, now uniformly distributed in $\{0, 1, \ldots, 63\}$. The state of the station then jumps to $(1, X)$ where the first component indicates that the station

---

[1]This additional Slot Time has a negligible impact on the results and is neglected in [17].

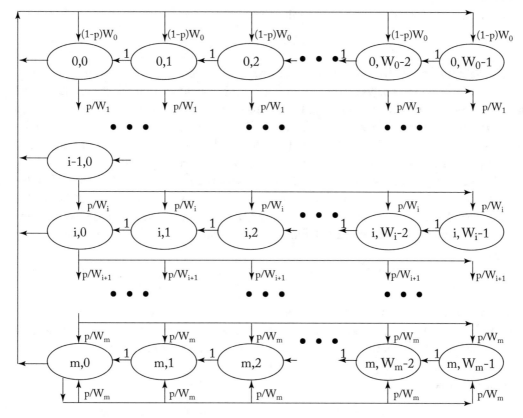

Figure 4.4: Transition probabilities for the backoff window size Markov chain.

already made one attempt. If the transmission succeeds, the state of the station jumps to the top of the diagram. The other states are explained in the same way.

The key simplifying assumption in this model is that, when the station transmits, it is successful with probability $1 - p$ and collides with probability p, independently of the state $(i, 0)$ of the station and of its previous sequence of states. This assumption makes it possible to analyze one station in isolation, and is validated using simulations. Since p is the probability of collision given that the station of interest transmits, we refer to it as the conditional collision probability. By analyzing the Markov chain, one can find the probability that its state is of the form $(i, 0)$ and that the station then attempts to transmit. Let us call this probability α. That is, α is the probability that at a given transmission opportunity (i.e., when its backoff counter reaches 0), this station attempts to transmit. Assume that all the n stations have the same average probability α of transmitting. Assume also that their transmission attempts are all independent. The probability $1 - p$ that this station succeeds (and does not collide) is then the probability that the other $n - 1$ stations do not transmit. That is,

$$1 - p = (1 - \alpha)^{n-1}.$$

Indeed, we explained in Section 3.10 that the probabilities of independent event multiply.

Summing up, the analysis proceeds as follows. One assumes that the station has a probability $1 - p$ of success when it transmits. Using this value of p, one solves the Markov chain and derives the attempt probability α for each station. This α depends on p, so that $\alpha = \alpha(p)$, some function of p. Finally, one solves $1 - p = (1 - \alpha(p))^{n-1}$. This is an equation in one unknown variable p. Having solved for p, we know the value of the attempt probability $\alpha(p)$.

We outline below how the Markov chain is solved to obtain $\alpha(p)$. Let π and P denote the invariant distribution and the probability transition matrix for the Markov chain, respectively. See the appendix for an introduction to these concepts. The transition probabilities corresponding to P are shown in Figure 4.4. By definition of the invariant distribution, we have $\pi = \pi P$. This gives

$$\pi(x) \sum_{y \neq x} P(x, y) = \sum_{y \neq x} \pi(y) P(y, x) \text{ for each } x.$$

This identity can be interpreted to say that the "total flow" in and out of a state are equal in equilibrium. These are referred to as the balanced equations for the Markov chain. For the Markov chain under consideration, recall that the state is the two-dimensional vector $\{s, b\}$ defined earlier. Applying the balanced equations and using the transition probabilities shown in Figure 4.4, it can be seen that

$$\pi(i - 1, 0)p = \pi(i, 0) \Rightarrow \pi(i, 0) = p^i \pi(0, 0) \text{ for } 0 < i < m. \tag{4.2}$$

Identity (4.2) is obtained by recursive applications of the balanced equations. Observe that, using the balanced equations $\pi(i - 1, 0)\frac{p}{W_i} = \pi(i, W_i - 1)$ and $\pi(i, W_i - 2) = \pi(i - 1, 0)\frac{p}{W_i} + \pi(i, W_i - 1)$, we can obtain $\pi(i, W_i - 2) = \pi(i - 1, 0)\frac{2p}{W_i}$. Continuing this way, we finally get, $\pi(i - 1, 0)p = \pi(i, 0)$. Applying the balanced equations recursively in a similar way and after some algebraic manipulations (see [17]), we get

$$\pi(m - 1, 0)p = (1 - p)\pi(m, 0) \Rightarrow \pi(m, 0) = \frac{p^m}{1 - p}\pi(0, 0), \text{ and} \tag{4.3}$$

$$\pi(i, k) = \frac{W_i - k}{W_i}\pi(i, 0), \text{ for } i \in \{0, \ldots, m\} \text{ and } k \in \{0, \ldots, W_i - 1\}. \tag{4.4}$$

Using the identities (4.2), (4.3), and (4.4), we find an expression for each $\pi(i, k)$ in terms of $\pi(0, 0)$, and then using the fact that $\sum_i \sum_k \pi(i, k) = 1$, we solve for $\pi(0, 0)$, and hence each $\pi(i, k)$. Recalling that

$$\alpha = \sum_{i=0}^{m} \pi(i, 0),$$

we finally find

$$\alpha(p) = \frac{2(1-2p)}{(1-2p)(W+1) + pW(1-(2p)^m)},$$

where $W = CW_{\min} + 1$.

Using α we calculate the network throughput (overall rate of data transmission) as follows. Time duration between two consecutive epochs of the Markov chain has a successful transmission with probability $\beta := n\alpha(1-\alpha)^{n-1}$. Let T be the average length of this duration. The network throughput is then given by $\beta B / T$ where B is the average number of data bits transmitted during a successful transmission. Indeed, during a typical duration, 0 data bits are transmitted with probability $1 - \beta$, and an average of B data bits are transmitted with probability β. Thus, βB is the average number of data bits transmitted in an average duration of T.

To calculate T, one observes that T corresponds to either an idle Slot Time or a duration that contains one or more simultaneous transmissions. If the duration contains exactly one transmission, T corresponds to the transmission time of a single packet, an *SIFS*, the transmission time of the ACK, and a *DIFS*. In the case of more than one transmission in this duration, T corresponds to the longest of the transmission times of the colliding packets and a *DIFS*. See [17] for the details regarding computation of T.

Simulations in [17] confirm the quality of the analytical approximation. That paper also extends the main analysis approach outlined here and presents extensive analysis for many other interesting aspects, including an analysis of theoretically maximum throughput achievable if the Congestion Window were not to be constrained by the standards specifications and were picked optimally depending on n.

Figure 4.5 shows the results obtained by solving the model described above with the parameters associated with 802.11b. We assume propagation delay is negligible, and that all WiFi data packets carry 1,500 bytes of data payload. As in [17], we also ignore the effect of ACK Time-Out and EIFS. As to be expected, the figure shows that the conditional collision probability increases as the number of devices increases. This causes the network throughput to decrease as the number of devices increases.

4.5   RECENT ADVANCES

4.5.1   IEEE 802.11N—INTRODUCTION OF MIMO IN WIFI

The IEEE 802.11n standard, published in 2009, supports a data rate of up to 600 Mbps. It makes use of Multiple-Input Multiple-Output (MIMO) communication in the unlicensed bands around both 2.4 GHz and 5 GHz, allowing simultaneous use of up to 4 spatially separated data streams. The number of simultaneous streams is clearly limited by the maximum number of transmit and receive antennas present at the two ends of the wireless link. Furthermore, it is also limited by the maximum number of simultaneous streams that can be supported by the radio hardware. The notation axb:c is used to indicate that the maximum number of transmit and receive antennas are a and b, respectively, and the maximum number of spatial streams

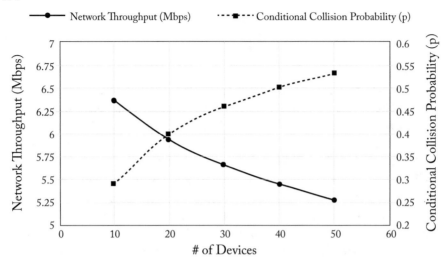

Figure 4.5: Network throughput and conditional collision probability.

allowed by the radio hardware is c. The IEEE 802.11n standard allows up to 4x4:4 configuration. It should be noted that the MIMO communication allowed by IEEE 802.11n is limited to a single pair of devices at any given time. This form of MIMO is referred to as Single-User MIMO (SU-MIMO). This restriction was subsequently eased with the IEEE 802.11ac standard discussed below. Besides MIMO, another Physical Layer enhancement in IEEE 802.11n is use of a larger channel width of 40 MHz. Additionally, in order to support a higher data rate, it uses the MAC layer feature called "frame aggregation" to allow multiple data frames to be aggregated in a single transmission, and thereby reduces the transmission overhead of headers and preamble.

4.5.2   IEEE 802.11AD—WIFI IN MILLIMETER WAVE SPECTRUM

The IEEE 802.11ad standard, published in 2012, is an amendment that expands for WiFi networks the use of the unlicensed band around 60 GHz. Since the radio waves in this band have their wavelengths close to 1 mm, they are referred to as the millimeter waves. These waves have very different transmission characteristics than the waves in 2.4 GHz and 5 GHz bands. They have much shorter range and are useful for only line-of-sight communication. Furthermore, they experience severe fading in rain. The IEEE 802.11ad standard makes use of large channel width of about 2 GHz, and offers a maximum data rate of 7 Gbps. The typical use cases for IEEE 802.11ad include the applications requiring high data rates over a short distance. They include wireless storage and display for desktop computers and in-home distribution of video. The technology enabling IEEE 802.11ad was originally developed and promoted by the Wireless Gigabit Alliance (WiGig), and this role was taken over later by the WiFi Alliance.

4.5.3 IEEE 802.11AC—INTRODUCTION OF MU-MIMO IN WIFI

The IEEE 802.11ac standard, published in 2013, is a significant evolution of the WiFi technology. It is specified for the unlicensed band around 5 GHz, and offers the maximum aggregate capacity of 6.8 Gbps, and maximum single user rate of 3.4 Gbps. One of the key advancements supported by the IEEE 802.11ac standard is Multi-User MIMO (MU-MIMO), where multiple users may simultaneously receive data from the AP. The standard only supports MU-MIMO in the downlink direction. Furthermore, the maximum number of user devices served simultaneously by an AP is limited to 4. Due to the difficult task of coordinating among multiple users, MU-MIMO in the uplink direction was deferred to a future standard (IEEE 802.11ax). MU-MIMO cannot only reduce the finer time-scale delay by reducing the need for time-multiplexing, it can also significantly improve network throughput, e.g., when instantaneous throughput is limited by the capabilities of user devices. In the notation discussed in the IEEE 802.11n subsection above, the IEEE 802.11ac standard allows up to 8x8:8 MIMO configuration. This enhanced configuration with the larger arrays of receive and transmit antennas and the increase in the number of simultaneous streams supported to 8 (from 4 by IEEE 802.11n) allows the network to derive the benefits of MU-MIMO more effectively. Another Physical Layer enhancement in IEEE 802.11ac is the larger channel width of up to 160 MHz, which may be created by bonding two contiguous or discontiguous 80 MHz channels. The higher data rates made possible by the IEEE 802.11ac standard make a variety of services, including video streaming, AP attached storage, and innovative cloud base services, practically feasible.

4.5.4 IEEE 802.11AH—WIFI FOR IOT AND M2M

The IEEE 802.11ah standard, published in 2016, operates in a sub-GHz unlicensed band. The exact frequency band used by this standard can be different in different countries. In the U.S., the unlicensed band around 900 MHz is used. Low power consumption and extended communication range are the key hallmarks of this standard. The lower frequency band used by this standard extends the communication range by reducing signal attenuation, and curtailing distortions due to walls and obstructions. As for the IEEE 802.11ac standard, the IEEE 802.11ah standard also supports downlink MU-MIMO. It supports up to 4x4:4 MIMO configuration with the maximum of 4 end devices served simultaneously by an AP. The targeted use cases include supporting Machine-to-Machine (M2M) scenarios, e.g., inter-connected actuators, Internet of Things (IoT) scenarios, e.g., sensors and smart meters, and offloading of traffic from the cellular networks. The standard specifies mandatory support for 1 MHz and 2 MHz channels for a low data rate starting from 150 Kbps, and optional support for up to 16 MHz wide channels for higher data rates up to 347 Mbps (aggregated over 4 spatial streams). To support IoT scenarios with a large number of stations in geographic proximity, the standard allows up to 8,192 stations to be associated with an AP.

Other notable features of IEEE 802.11ah include support for sectorization, Restricted Access Window (RAW), Target Wakeup Time (TWT), and relay nodes. In order to mitigate the hidden terminal problem in the presence of a large number of low power stations, the IEEE 802.11ah standard makes use of group sectorization based on the station locations. RAW and TWT limit the time when an end device can transmit. TWT further restricts this time according to the assigned transmission schedule. RAW and TWT help reduce channel contention. Furthermore, TWT in particular allows an end device to save power by going to sleep until its expected wakeup time. An end device, referred to as a Relay Access Point (RAP), can have the capability to act as a relay node to forward a packet to an Access Point. This helps end devices to consume to less power to transmit a packet when there is an RAP relatively closer than the AP. On the other hand, multi-hopping has the downside of additional latency and more usage of the wireless medium. To find a useful compromise, the IEEE 802.11ah allows use of at most one RAP on the way to the AP.

The WiFi Alliance recently announced a power-efficient solution, referred to as WiFi HaLow, based on the IEEE 802.11ah standard to address use cases needing low power consumption at the end devices. Examples of such use cases include connected cars, smart home applications, and healthcare devices.

4.5.5   PEER-TO-PEER WIFI

It is convenient to be able to transfer files between two or more devices without having to rely on a traditional AP, e.g., for printing, for sharing pictures, audio, or video, or for communicating with a "hotspot" device in order to share its Internet access. There are two basic ways for devices to connect with each other over WiFi without requiring the presence of a traditional AP. One possibility is for one of the devices to emulate the AP functions. This can be done with help from advanced WiFi chipsets and/or by emulating the AP functions in software. For example, a solution referred to as WiFi Direct, promoted by an industrial consortium called WiFi Alliance, takes this approach. There are also proprietary solutions based on the same principle. If access to the Internet is desired, it's only permitted by sharing the Internet access of the device emulating the AP functions. The overall solution has to specify how the election of the device acting like an AP is done, and what happens if that device leaves or fails. In an alternate approach, a group of WiFi devices can establish an ad hoc network among themselves where each pair of devices has a peer-to-peer WiFi link. A standardized way for forming and operating such an ad hoc network is included in the IEEE 802.11 standards.

4.6   APPENDIX: MARKOV CHAINS

A *Markov chain* describes the state X_n at time $n = 0, 1, 2 \ldots$ of a system that evolves randomly in a finite set $\mathcal{X}$ called the state space. (The case of an infinite state space is more complex.) The rules of evolution are specified by a matrix $P = [P(i, j), i, j \in \mathcal{X}]$ called the transition probability matrix. This matrix has nonnegative entries and its rows sum to one. That is,

$$P(i, j) \geq 0, i, j \in \mathcal{X} \text{ and } \sum_j P(i, j) = 1, i \in \mathcal{X}.$$

When the state is i, it jumps to j with probability $P(i, j)$, independently of its previous values. That is,

$$P[X_{n+1} = j | X_n = i, X_m, m < n] = P(i, j), i, j \in \mathcal{X}, n = 0, 1, 2, \ldots.$$

This expression says that the probability that $X_{n+1} = j$ given that $X_n = i$ and given the previous values $\{X_m, m < n\}$ is equal to $P(i, j)$. You can imagine that if $X_n = i$, then the Markov chain rolls a die to decide where to go next. The die selects the value j with probability $P(i, j)$. The probabilities of the possible different values add up to one.

Figure 4.6 illustrates two Markov chains with state space $\mathcal{X} = \{1, 2, 3\}$. The one on the left corresponds to the transition probability matrix P given below:

$$P = \begin{bmatrix} 0 & 1 & 0 \\ 0.6 & 0 & 0.4 \\ 0.3 & 0 & 0.7 \end{bmatrix}.$$

The first row of this matrix states that $P(1, 1) = 0, P(1, 2) = 1, P(1, 3) = 0$, which is consistent with the diagram in Figure 4.6. In that diagram, an arrow from i to j is marked with $P(i, j)$. If $P(i, j) = 0$, then there is no arrow from i to j. The diagram is called the *state transition diagram* of the Markov chain.

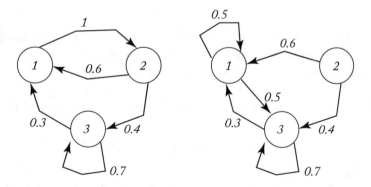

Figure 4.6: Two Markov chains: irreducible (left) and reducible (right).

The two Markov chains are similar, but they differ in a crucial property. The Markov chain on the left can go from any i to any j, possibly in multiple steps. For instance, it can go from 1–3 in two steps: from 1–2, then from 2–3. The Markov chain on the right cannot go from 1–2. We say that the Markov chain on the left is *irreducible*. The one on the right is *reducible*.

Assume that the Markov chain on the left starts at time 0 in state 1 with probability $\pi(1)$, in state 2 with probability $\pi(2)$, and in state 3 with probability $\pi(3)$. For instance, say that $\pi = [\pi(1), \pi(2), \pi(3)] = [0.3, 0.2, 0.5]$. What is the probability that $X_1 = 1$? We find that probability as follows:

$$P(X_1 = 1) = P(X_0 = 1)P(1, 1) + P(X_0 = 2)P(2, 1) + P(X_0 = 3)P(3, 1)$$
$$= 0.3 \times 0 + 0.2 \times 0.6 + 0.5 \times 0.3 = 0.27. \tag{4.5}$$

Indeed, there are three exclusive ways that $X_1 = 1$: Either $X_0 = 1$ and the Markov chain jumps from 1–1 at time 1, or $X_0 = 2$ and the Markov chain jumps from 2–1 at time 1, or $X_0 = 3$ and the Markov chain jumps from 3–1 at time 1. The identity above expresses that the probability that $X_1 = 1$ is the sum of the probabilities of the three exclusive events that make up the event that $X_1 = 1$. This is consistent with the fact that the probabilities of exclusive events add up, as we explained in Section 3.10.2.

More generally, if π is the (row) vector of probabilities of the different states at time 0, then the row vector of probabilities of the states at time 1 is given by πP, the product of the row vector π by the matrix P. For instance, the expression (4.5) is the first component of πP.

In our example, $\pi P \neq \pi$. For instance, $P(X_1 = 1) = 0.27 \neq P(X_0 = 1) = 0.3$. That is, the probabilities of the different states change over time. However, if one were to start the Markov chain at time 0 with probabilities π such that $\pi P = \pi$, then the probabilities of the states would not change. We call such a vector of probabilities an *invariant* distribution. The following result is all we need to know about Markov chains.

Theorem 4.1

Let X_n be an irreducible Markov chain on a finite state space $\mathcal{X}$ with transition probability matrix P. The Markov chain has a unique invariant distribution π. Moreover, the long-term fraction of time that $X_n = i$ is $\pi(i)$ for $i \in \mathcal{X}$, independently of the distribution of X_0.

For our Markov chain on the left of Figure 4.6, we find the invariant distribution π by solving

$$\pi P = \pi \text{ and } \pi(1) + \pi(2) + \pi(3) = 1.$$

After some algebra, one finds $\pi = [0.3, 0.3, 0.4]$.

The long-term fraction of time that $X_n = 1$ is then equal to 0.3, and this fraction of time does not depend on how one starts the Markov chain.

The Markov chain on the right of Figure 4.6 is reducible, so the theorem does not apply. However, it is quite clear that after a few steps, the Markov chain ends up being in the subset $\{1, 3\}$ of the state space. One can then consider the Markov chain reduced to that subset, and it is now irreducible. Applying the theorem, we find that there is a unique invariant distribution, and we can calculate it to be $[3/8, 0, 5/8]$.

4.7   SUMMARY

- A WiFi BSS is a set of devices communicating over a common channel using the WiFi MAC protocol. We focus on the infrastructure mode WiFi networks where all communication takes place via the AP.

- WiFi networks operate in an unlicensed band around 2.4 or 5 GHz. In the 2.4 GHz band, they typically operate over channels 1, 6, and 11.

- IEEE 802.11, 802.11a, 80211b, 802.11g, 802.11n, 802.11ad, 802.11ac, and 802.11ah are the key specifications with different Physical Layers for WiFi networks. These specifications support Physical Layer rates up to 7 Gbps.

- The MAC Sublayer in WiFi networks is based on the CSMA/CA techniques. DCF is its prevalent mode of operation.

- Medium access is regulated using different IFS parameters, i.e., SIFS, DIFS, and EIFS.

- The MAC header of a data frame contains four MAC addresses. In a BSS with the AP having integrated router functions, only two of these addresses are relevant.

- RTS/CTS mechanism resolves the hidden terminal problem.

- Virtual Carrier Sensing using NAV is an important MAC Sublayer enhancement for avoiding collisions.

- Given the high MAC Sublayer and Physical Layer overheads, network efficiency (equivalently, system throughput) is an important characterization for WiFi networks.

- MIMO is a key enabling technology for achieving high throughput and reliability. Starting with IEEE 802.11ac, MU-MIMO make MIMO even more effective for offering higher performance.

- IEEE 802.11ah is a sub-GHz WiFi specification. IoT and M2M are among its key use cases.

- WiFi Direct is a solution for P2P networking using the idea of Soft AP.

4.8   PROBLEMS

P4.1 Consider a wireless network shaped like a pentagon. The wireless nodes are shown at the vertices A, B, C, D, and E, and the nodes are placed such that each node can talk only to its two neighbors—as shown. Thus there are 10 unidirectional wireless links in this network. Assume that the nodes employ RTS/CTS and also require ACKs for successful transmission.

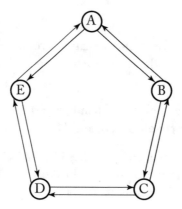

Figure 4.7: Figure for WiFi Problem 1.

Consider a situation when A is transmitting a packet to B. Obviously, link A→B is active, and all links that are affected by this transmission must keep quiet. Considering RTS/CTS, and ACKs, indicate which of the other links could also be active at the same time. In other words, indicate which of the other links could be simultaneously transmitting.

P4.2 Consider a Wireless LAN (WLAN) operating at 11 Mbps that follows the 802.11 MAC protocol with the parameters described below. All the data frames are of a fixed size. Assume that the length of the average contention window (after DIFS following last transmission, time elapsed until next transmission) is 1.5T Slot Times where T is the transmission time (in Slot Times), including the overhead, of the frame contending for medium access. Assume the parameters for this WLAN as follows: Slot Time = 20 μs, DIFS = 2.5 Slot Times, SIFS = 0.5 Slot Times, RTS = CTS = ACK = 10 Slot Times (including overhead), and data frame overhead = 9 Slot Times.

Determine the threshold value (in bytes) for the data payload such that the data transfer efficiency is greater with the RTS/CTS mechanism for frames with data payload larger than this threshold.

P4.3 Suppose there are only two nodes, a source and a destination, equipped with 802.11b wireless LAN radios that are configured to use RTS/CTS for packets of all sizes. The source node has a large amount of data to send to the destination node. Also, the nodes are separated by 750 m and have powerful enough radios to communicate at this range. No other nodes operate in the area. What will be the data throughput between source and destination assuming that packets carry 1,100 bytes of data?

Assume the following parameters:

  • Propagation speed is 3×10^8 m/s.

- Slot Time $= 20\,\mu s$, SIFS $= 10\,\mu s$, DIFS $= 50\,\mu s$, and $CW_{min} = 31$ Slot Times.
- The preamble, the Physical Layer header, the MAC header, and trailer take a combined $200\,\mu s$ per packet to transmit.
- The Data is transmitted at 11 Mbps.
- ACK, RTS, and CTS each take $200\,\mu s$ to transmit.

P4.4 Consider the Markov chain model due to G. Bianchi [17] discussed in the chapter for the scenario where only the Access Point and a single WiFi device have unlimited data to send to each other. Assume that $CW_{min} = CW_{max} = 1$ and Slot Time $= 20\,\mu s$.

  (a) Draw the transition probability diagram for the Markov chain.

  (b) Derive the invariant distribution for the Markov chain.

  (c) Calculate the probability that a duration between two consecutive epochs of the Markov chain would have a successful transmission.

  (d) Assume that each WiFi data frame transports 1,000 bytes of payload. Furthermore, assume that a duration between two consecutive Markov chain epochs with a successful transmission is 60 Slot Time long, while that with a collision is 50 Slot Time long. Calculate the network throughput in Mbps as seen by the upper layers.

P4.5 See P7.7 of Transport Chapter. It is on use of UDP over WiFi.

P4.6 See P7.8 of Transport Chapter. It is on use of TCP over WiFi.

4.9   REFERENCES

[41] has a good summary of the evolution of the IEEE 802.11 standards. The detailed specifications of the IEEE 802.11 standards can be found in [42, 43, 44, 45, 46, 47, 48]. The text by Gast [35] describes the basics of the IEEE 802.11 standards. The efficiency model is due to Bianchi [17]. [13, 55, 98] discuss various aspects of the IEEE 802.11ah standard. Further information about WiFi Direct can be found in [23, 106]. For details on Markov chains, see [39] or [104].

CHAPTER 5

Routing

This chapter explores how various networks determine the paths that packets should follow.

Internet routers use a two-level scheme: inter-domain and intra-domain routing. The intra-domain algorithms find the shortest path to the destination. These mechanisms extend easily to "anycast" routing when a source wants to reach any one destination in a given set. We also discuss multicast routing, which delivers packets to all the destinations in a set. The inter-domain routing uses an algorithm where each domain selects a path to a destination domain based on preference policies.

We conclude with a discussion of routing in ad hoc wireless networks.

5.1 DOMAINS AND TWO-LEVEL ROUTING

The nodes in the Internet are grouped into about 55,000 (in 2017) *Autonomous Systems* (*domains*) that are under the management of separate entities. For instance, the Berkeley campus network is one domain, so is the MIT network, and so are the AT&T, Verizon, and Sprint cross-country networks. The routing in each domain is performed by a shortest-path protocol called an *intra-domain routing protocol*. The routing across domains is implemented by an *inter-domain routing protocol*. Most domains have only a few routers. A few large domains have hundreds or even thousands of routers. A typical path in the Internet goes through less than half a dozen domains. The Internet has a *small-world* topology where two domains are only a few hand-shakes away from each other, like we all are.

There are two reasons for this decomposition. The first one is scalability. Intra-domain shortest path routing requires a detailed view of the domain; such a detailed view is not possible for the full Internet. The second reason is that domain administrators may prefer—for economic, reliability, security, or other reasons—to send traffic through some domains rather than others. Such preferences require a different type of protocol than the strict shortest path intra-domain protocols.

5.1.1 SCALABILITY

The decomposition of the routing into two levels greatly simplifies the mechanisms. To appreciate that simplification, say that finding a shortest path between a pair of nodes in a network with N nodes requires each node to send about N messages to the other nodes. In that case, if a network consists of M domains of N nodes each, a one-level shortest path algorithm requires each node sending about MN messages. On the other hand, in a two-level scheme, each node sends N messages to the other nodes in its domain and one representative node in each domain may send M messages to the representatives in the other domains.

5.1.2   TRANSIT AND PEERING

The different *Internet Service Providers* (ISPs) who own networks make agreements to carry each other's traffic. There are two types of agreement: peering and transit. In a *peering* agreement, the ISPs reciprocally provide free connectivity to each other's local or "inherited" customers. In a *transit* agreement, one ISP provides (usually sells) access to all destinations in its routing table.

Figure 5.1 illustrates these agreements. The left part of the figure shows three ISPs (ISP A, ISP B, ISP C). ISP A and ISP B have a peering agreement. Under this agreement, ISP A announces the routes to its customers ($A_1, ..., A_k$). Similarly, ISP B announces the routes to its customers ($B_1, ..., B_m$). The situation is similar for the peering agreement between ISP B and ISP C. Note, however, that ISP B does not provide connectivity between ISP A and ISP C.

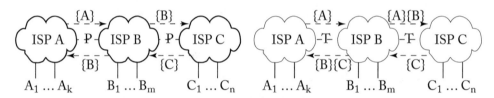

Figure 5.1: Peering (left) and transit (right) agreements.

The right part of Figure 5.1 shows transit agreements between ISP A and ISP B and between ISP B and ISP C. Under this agreement, ISP B announces the routes to all the customers it is connected to and agrees to provide connectivity to all those customers. Thus, ISP B provides connectivity between ISP A and ISP C (for a fee).

Figure 5.2 shows a fairly typical situation. Imagine that ISP A and ISP B are two campuses of some university. ISP C is the ISP that provides connectivity between the campuses and the rest of the Internet, as specified by the transit agreement between ISP C and the two campuses. By entering in a peering agreement, ISP A and ISP B provide direct connectivity between each other without having to go through ISP C, thus reducing their Internet access cost. Note that the connection between ISP A and ISP C does not carry traffic from ISP B, so that campus A does not pay transit fees for campus B.

5.2   INTER-DOMAIN ROUTING

At the inter-domain level, each domain essentially looks like a single node. The inter-domain routing problem is to choose a path across those nodes to go from one source domain to a destination domain. For instance, the problem is to choose a sequence of domains to go from the Berkeley domain to the MIT domain. A natural choice would be the path with the fewest

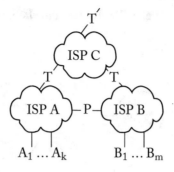

Figure 5.2: Typical agreements.

hops. However, Berkeley might prefer to send its traffic through the Sprint domain because that domain is cheaper or more reliable than the AT&T domain.[1]

For that reason, the inter-domain protocol of the Internet—currently the *Border Gateway Protocol*, or BGP, is based on a path vector algorithm.

5.2.1 PATH VECTOR ALGORITHM

The most deployed inter-domain routing protocol in the Internet is the *Border Gateway Protocol* which is based on a path vector algorithm. When using a *path vector algorithm*, as shown in Figure 5.3, the routers advertise to their neighbors their preferred path to the destination. The figure

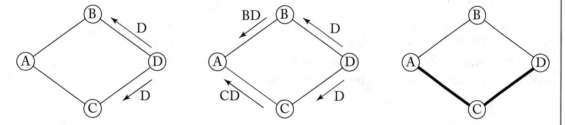

Figure 5.3: Path vector algorithm.

again considers destination D. Router D advertises to its neighbors B and C that its preferred path to D is D. Router B then advertises to A that its preferred path to D is BD; similarly, C advertises to A that its preferred path to D is CD. Router A then selects its preferred path to D among ABD and ACD. Since the full paths are specified, router A can base its preference not only on the number of hops but also on other factors, such as existing transit or peering agreements, pricing, security, reliability, and various other considerations. In the figure, we assume

[1]This situation is hypothetical. All resemblance to actual events is purely accidental and unintentional.

that router A prefers the path ACD to ABD. In general, the preferred path does not have to be the shortest one.

Figure 5.4 provides an illustration.

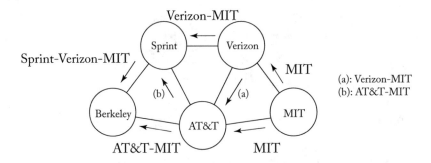

Figure 5.4: Routing across domains: inter-domain routing.

The inter-domain protocol of the Internet enables network managers to specify preferences other than shortest path. Using that protocol, the Berkeley domain is presented with two paths to the MIT domain: Sprint-Verizon-MIT and AT&T-MIT. The Berkeley manager can specify a rule that states, for instance: (1) If possible, find a path that does not use AT&T; (2) Among the remaining paths, choose the one with the fewest hops; (3) In case of a tie, choose the next domain in alphabetical order. With these rules, the protocol would choose the path Sprint-Verizon-MIT. (In the actual protocol, the domains are represented by numbers, not by names.)

Summarizing, each domain has one or more representative routers and the representatives implement BGP. Each representative has a set of policies that determine how it selects the preferred path among advertised paths. These policies also specify which paths it should advertise, as explained in the discussion on peering and transit.

5.2.2   POSSIBLE OSCILLATIONS

Whereas it is fairly clear that one can find a shortest path without too much trouble, possibly after breaking ties, it is not entirely clear that a path vector protocol converges if domain administrators select arbitrary preference rules. In fact, simple examples show that some rules may lead to a lack of convergence. Figure 5.5 illustrates such an example. The figure shows four nodes A, B, C, D that represent domains. The nodes are fully connected. The preferences of the nodes for paths leading to D are indicated in the figure. For instance, node A prefers the path ACD to the direct path AD. Similarly, node C prefers CBD to CD, and node B prefers BAD to BD. For instance, say that the links to D are slower than the other links and that the preference is to go counter-clockwise to a faster next link. Assume also that nodes prefer to avoid three-hop paths. The left part of the figure shows that A advertises its preferred path ACD to node B. Node B

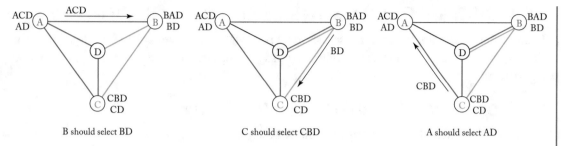

Figure 5.5: A path vector protocol may fail to converge.

then sees the path *ACD* advertised by *A*, the direct path *BD*, and the path *CBD* that *C* advertises. Among these paths, *B* selects the path *BD* since *BACD* would have three hops and going to *C* would induce a loop. The middle part of the figure shows that *B* advertises its preferred path *BD* to node *C*. Given its preferences, node *C* then selects the path *CBD* that it advertises to node *A*, as shown in the right part of the figure. Thus, the same steps repeat and the choices of the nodes keep on changing at successive steps. For instance, *A* started preferring *ACD*, then it prefers *AD*, and so on. This algorithm fails to converge.

The point of this example is that, although path vector enables domain administrators to specify preferences, this flexibility may result in poorly behaving algorithms.

5.2.3   MULTI-EXIT DISCRIMINATORS

Figure 5.6 shows ISP A attached to ISP B by two connections. ISP A informs ISP B that the top connection is closer to the destinations X than the bottom connection. To provide that information, ISP A attaches a *discriminator* to its local destinations. The discriminator represents a metric from the router attached to the connection to the destination. With this information the router R4 in ISP B calculates that it should send traffic destined for X via the upper connection. ISP B does not forward these discriminators.

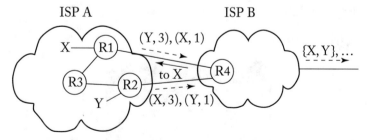

Figure 5.6: Multi-exit discriminators.

5.3   INTRA-DOMAIN SHORTEST PATH ROUTING

Inside a single domain, the routers use a shortest path algorithm. We explain two algorithms for finding shortest paths: Dijkstra and Bellman-Ford. We also explain the protocols that use those algorithms.

5.3.1   DIJKSTRA'S ALGORITHM AND LINK STATE

Figure 5.7 illustrates the operations of a link state algorithm.

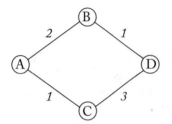

(1) Exchange Link States
   A: [B, 2], [C, 1]
   B: [A, 2], [D, 1]
   C: [A, 1], [D, 3]
   D: [B, 1], [C, 3]

(2) Each router computes the shortest paths to the other routers and enters the results in its routing table.

Figure 5.7: Link state routing algorithm.

When using a *link state algorithm*, the routers exchange link state messages. The link state message of one router contains a list of items [*neighbor, distance to neighbor*]. That list specifies the metric of the link to each neighbor. The metric is representative of the time the node takes to send a message to that neighbor. In the simplest version, this metric is equal to one for all the links. In a more involved implementation, the metric is a smaller number for faster links. The length of a path is defined as the sum of the metrics of the links of that path.

Thus, every router sends a link state message to every other router. After that exchange, each router has a complete view of the network, and it can calculate the shortest path from each node to every other node in the network. The next hop along the shortest path to a destination depends only on the destination and not on the source of the packet. Accordingly, the router can enter the shortest paths in its routing table. For instance, the routing table of node A contains the information shown in Table 5.1.

Table 5.1: Routing table of router A

Destination	Next Hop
B	B
C	C
D	B

The top-left graph in Figure 5.8 shows nodes attached with links; the numbers on the links represent their metric. The problem is to find the shortest path between pairs of nodes.

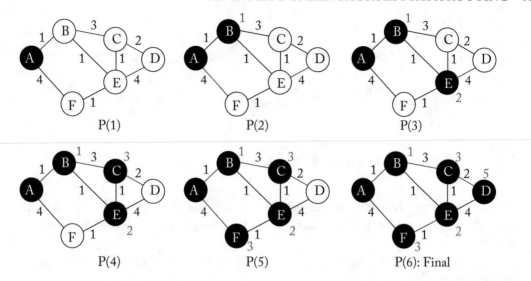

Figure 5.8: Dijkstra's routing algorithm.

To find the shortest paths from node A to all the other nodes, Dijkstra's algorithm computes recursively the set $P(k)$ of the k-closest nodes to node A (shown with black nodes). The algorithm starts with $P(1) = \{A\}$. To find $P(2)$, it adds the closest node to the set $P(1)$, here B, and writes its distance to A, here 1. To find $P(3)$, the algorithm adds the closest node to $P(2)$, here E, and writes its distance to A, here 2. The algorithm continues in this way, breaking ties according to a deterministic rule, here favoring the lexicographic order, thus adding C before F. This is a one-pass algorithm whose number of steps is the number of nodes.

After running the algorithm, each node remembers the next step along the shortest path to each node and stores that information in a routing table. For instance, A's routing table specifies that the shortest path to E goes first along link AB. Accordingly, when A gets a packet destined to E, it sends it along link AB. Node B then forwards the packet to E.

5.3.2 BELLMAN-FORD AND DISTANCE VECTOR

Figure 5.9 shows the operations of a distance vector algorithm. The routers regularly send to their neighbors their current estimate of the shortest distance to the other routers. The figure only considers destination D, for simplicity. Initially, only node D knows that its distance to D is equal to zero, and it sends that estimate to its neighbors B and C. Router B adds the length 1 of the link BD to the estimate 0 it got from D and concludes that its estimated distance to D is 1. Similarly, C estimates the distance to D to be equal to $3 + 0 = 3$. Routers B and C send those estimates to A. Node A then compares the distances $2 + 1$ of going to D via node B and $1 + 3$ that corresponds to going to C first. Node A concludes that the estimated shortest distance to

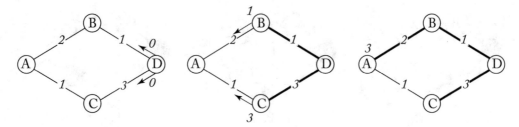

Figure 5.9: Distance vector routing algorithm.

D is 3. At each step, the routers remember the next hop that achieves the shortest distance and indicates that information in their routing table. The figure shows these shortest paths as thick lines.

For instance, the routing table at router A is the same as using the link state algorithm and is shown in Table 5.1.

Bad News Travel Slowly

One difficulty with this algorithm is that it may take a long time to converge to new shortest paths when a link fails, as the example below shows. The top of Figure 5.10 shows three nodes A, B, C (on the right) attached with links with metric 1. The Bellman-Ford algorithm has converged to the estimates of the lengths of the shortest paths to destination C. At that point, the link between nodes B and C breaks down.

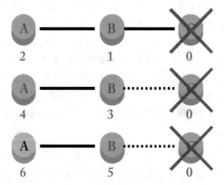

Figure 5.10: Bad news travel slowly when nodes use the Bellman-Ford algorithm.

The middle part of the figure shows the next iteration of the Bellman-Ford algorithm. Node A advertises its estimate 2 of the shortest distance to node C, and node B realizes that the length of the link BC is now infinite. Node B calculates its new estimate of the shortest distance to C as $1 + 2 = 3$, which is the length of the link from B to A plus the distance 2 that A advertises from A to C. Upon receiving that estimate 3 from B, A updates its estimate to $1 + 3 = 4$. The bottom part of the figure shows the next iteration of the algorithm. In the subsequent iterations, the estimates keep increasing.

We explain how to handle such slow convergence in the next section.

Convergence

Assume that the graph does not change and that the nodes use the Bellman-Ford algorithm. More precisely, say that at step $n \geq 0$ of the algorithm each node i has an estimate $d_n(i)$ of its shortest distance to some fixed destination i_0. Initially, $d_0(i) = \infty$ for $i \neq i_0$ and $d_0(i_0) = 0$. At step $n + 1$, some node i receives the estimate $d_n(j)$ from one of its neighbors j. At that time, node i updates its estimate as

$$d_{n+1}(i) = \min\{d_n(i), d(i, j) + d_n(j)\}.$$

The order in which the nodes send messages is arbitrary. However, one assumes that each node i keeps sending messages to every one of its neighbors.

The claim is that if the graph does not change, then $d_n(i) \to d(i)$, where $d(i)$ is the minimum distance from node i to node i_0. To see why this is true, observe that $d_n(i)$ can only decrease after i has received one message. Also, as soon as messages have been sent along the shortest path from i_0 to i, $d_n(i) = d(i)$.

Note that this algorithm does not converge to the true values if it starts with different initial values. For instance, if it starts with $d_n(i) = 0$ for all i, then the estimates do not change. Consequently, for the estimates to converge again after some link breaks or increases its length, the algorithm has to be modified. One approach is as follows. Assume that node i receives an estimate from neighbor j that is larger than a previous estimate it got from that node previously. Then node i resets its estimate to ∞ and send that estimate to its neighbors. In this way, all the nodes reset their estimates and the algorithm restarts with $d_0(i) = \infty$ for $i \neq i_0$ and $d_0(i_0) = 0$. The algorithm then converges, as we saw earlier.

5.4   ANYCAST, MULTICAST

When a source sends a message to a single destination, we say that the message is *unicast*. In some applications, one may want to send a message to any node in a set. For instance, one may need to find one of many servers that have the same information or to contact any member of a group. In such a case, one says that the message must be *anycast*. In other applications, all members of a group must get the message. This is called *multicast*. In particular, if the message must be sent to all other nodes, we say that it is *broadcast*. We explain the anycast and multicast routing.

5.4.1   ANYCAST

Figure 5.11 shows a graph with two shaded nodes. The anycast routing problem is to find the

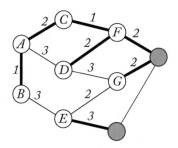

Figure 5.11: Shortest paths (thick lines) to set of shaded nodes.

shortest path from every other node to *any one* of the shaded nodes. One algorithm is identical to the Bellman-Ford algorithm for unicast routing. For a node i, let $d(i)$ be the minimum distance from that node to the anycast set. That is, $d(i)$ is the minimum distance to any node in the anycast set. If we denote by $d(i, j)$ the metric of link (i, j), then it is clear that

$$d(i) = \min\{d(i, j) + d(j)\}$$

where the minimum is over the neighbors j of node i. This is precisely the same relation as for unicast. The only difference is the initial condition: for anycast, $d(k) = 0$ for all node k in the anycast set.

Another algorithm is Dijkstra's shortest path algorithm that stops when it reaches one of the shaded nodes. Recall that this algorithm calculates the shortest paths from any given node to all the other nodes, in order of increasing path length. When it first finds one of the target nodes, the algorithm has calculated the shortest path from the given node to the set.

5.4.2   MULTICAST

Figure 5.12 shows a graph. Messages from A must be multicast to the two shaded nodes. The left part of the figure shows a tree rooted at A whose leaves are the shaded nodes and whose sum of link lengths is minimized over all such trees. This tree is called the *Steiner tree* (named after Jacob Steiner). The right part of the figure shows the tree of shortest paths from A to the shaded nodes. Note that the sum of the lengths of the links of the tree on the right (6) is larger than the sum of the lengths of the links of the Steiner tree (5).

Finding a Steiner tree is NP-hard. In practice, one uses approximations. One multicast protocol for Internet uses the tree of shortest paths.

The most efficient approach is to use one Steiner multicast tree for each multicast source. In practice, one may use a *shared tree* for different sources. For instance, one could build a Steiner from San Francisco to the major metropolitan areas in the U.S. If the source of a multicast is in Oakland, it can connect to that tree. This is less optimal than a Steiner tree from Oakland, but the algorithm is much less complex.

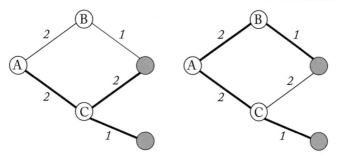

Figure 5.12: Minimum weight (Steiner) tree (left) and tree of shortest paths from A (right).

5.4.3 FORWARD ERROR CORRECTION

Using retransmissions to make multicast reliable is not practical. Imagine multicasting a file or a video stream to 1,000 users. If one packet gets lost on a link of the multicast tree, hundreds of users may miss that packet. It is not practical for them all to send a negative acknowledgment. It is not feasible either for the source to keep track of positive acknowledgments from all the users.

A simpler method is to add additional packets to make the transmission reliable. This scheme is called a *packet erasure code* because it is designed to be able to recover from "erasures" of packets, that is, from packets being lost in the network. For instance, say that you want to send packets $P1$ and $P2$ of 1 KByte to a user, but that it is likely that one of the two packets could get lost. To improve reliability, one can send $\{P1, P2, C\}$ where C is the addition bit by bit, modulo 2, of the packets $P1$ and $P2$. If the user gets any two of the packets $\{P1, P2, C\}$, it can recover the packets $P1$ and $P2$. For instance, if the user gets $P1$ and C, it can reconstruct packet P_2 by adding $P1$ and C bit by bit, modulo 2.

This idea extends to n packets $\{P1, P2, \ldots, Pn\}$ as follows: one calculates each of the packets $\{C1, C2, \ldots, Cm\}$ as the sum bit by bit, modulo 2, of a randomly selected set of packets in $\{P1, P2, \ldots, Pn\}$. The header of packet Ck specifies the subset of $\{1, 2, \ldots, n\}$ that was used to calculate Ck. If m is sufficiently large, one can recover the original packets from any n of the packets $\{C1, C2, \ldots, Cm\}$.

One decoding algorithm is very simple. It proceeds as follows:

- If one of the Ck, say Cj, is equal to one of the packets $\{P1, \ldots, Pn\}$, say Pi, then Pi has been recovered. One then adds Pi to all the packets Cr that used that packet in their calculation.

- One removes the packet Cj from the collection and one repeats the procedure.

- If at one step one does not find a packet Cj that involves only one Pi, the procedure fails.

Figure 5.13 illustrates the procedure. The packets $\{P1, \ldots, P4\}$ are to be sent. One calculates the packets $\{C1, \ldots, C7\}$ as indicated in the figure. The meaning of the graph is that a packet Cj is the sum of the packets Pi it is attached to, bit by bit, modulo 2. For instance,

$$C1 = P1, C2 = P1 + P2, C3 = P1 + P2 + P3,$$

and so on.

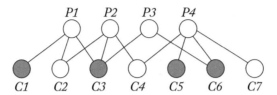

Figure 5.13: Calculation of FEC packets.

Now assume that the packets $C1, C3, C5, C6$ (shown shaded in Figure 5.13) are received by the destination. Following the algorithm described above, one first looks for a packet Cj that is equal to one of the Pi. Here, one sees that $C1 = P1$. One then adds $P1$ to all the packets Cj that used that packet. Thus, one replaces $C3$ by $C3 + C1$. One removes $P1$ from the graph and one is left with the graph shown in the left part of Figure 5.14.

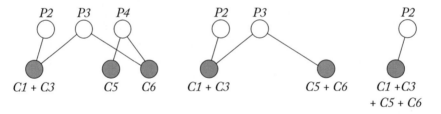

Figure 5.14: Updates in decoding algorithm.

The algorithm then continues as shown in the right part of the figure. Summing up, one finds successively that $P1 = C1$, $P4 = C5$, $P3 = C5 + C6$, and $P2 = C1 + C3 + C5 + C6$.

In practice, one chooses a distribution on the number D of packets Pi that are used to calculate each Cj. For instance, say that D is equally likely to be $1, 2$, or 3. To calculate $C1$, one generates the random variable D with the selected distribution. One then picks D packets randomly from $\{P1, \ldots, Pn\}$ and one adds them up bit by bit, modulo 2. One repeats the procedure to calculate $C2, \ldots, Cm$. Simulations show that with $m \approx 1.05 \times n$, the algorithm has a good likelihood to recover the original packets from any n of the packets $\{C1, \ldots, Cm\}$. For instance, if $n = 1,000$, one needs to send $1,050$ packets so that any $1,000$ of these packets suffice to recover the original $1,000$ packets, with a high probability.

5.4.4   NETWORK CODING

Network coding is an in-network processing method that can increase the throughput of multicast transmissions in a network. We explain that possibility in Figure 5.15.

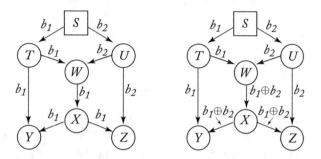

Figure 5.15: Multicast from S to X and Y without (left) and with (right) network coding.

In the Figure 5.15, source S sends packets to both Y and Z. b_1 and b_2 represent the bits to be transmitted. Say that the rate of every link in the network is R packets per second. The figure on the left has two parallel links from W to X. By using the coding in the intermediate node, as shown in the right part of the figure, the network can deliver packets at the rate of $2R$ to both Y and Z. Without network coding, as shown on the left, the network can only deliver one bit to Y and two bits to Z (or, two bits to Y and one bit to Z). One can show that the network can achieve a throughput of $1.5R$ to both Y and Z without network coding.

The general result about network coding and multicasting is as follows. Consider a general network and assume that the feasible rate from S to any one node in $\mathcal{N}$ is at least R. Then, using network coding, it is possible to multicast the packets from S to $\mathcal{N}$ at rate R. For instance, in the network of Figure 5.15, the rate from S to Y is equal to $2R$ as we can see by considering the two disjoint paths STY and $SUWXY$. Similarly, the rate from S to Z is also $2R$. Thus, one can multicast packets from S to Y and Z at rate $2R$.

Network coding can also be useful in wireless networks, as the following simple example shows. Consider the network in Figure 5.16. There are two WiFi clients X and Y that communicate via access point Z. Assume that X needs to send packet A to Y and that Y needs to send packet B to X.

The normal procedure for X and Y to exchange A and B requires transmitting four packets over the wireless channel: packet A from X to Z, packet A from Z to Y, packet B from Y to Z, and packet B from Z to X. Using network coding, the devices transmit only three packets: packet A from X to Z, packet B from Y to Z, and packet $A \oplus B$ (the bit by addition, modulo 2, of A and B) broadcast from Z to X and Y. Indeed, X knows A so that when it gets $C = A \oplus B$, it can recover B by calculating $C \oplus A$. Similarly, Y gets A as $A = C \oplus B$.

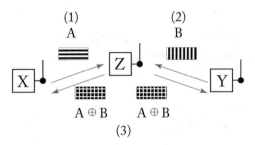

Figure 5.16: Network coding for a WiFi network.

This saving of one packet out of four requires delaying transmissions and keeping copies. Also, the savings are smaller if the traffic between X and Y is not symmetric. Thus, although this observation is cute, its benefits may not be worth the trouble in a real network.

5.5   AD HOC NETWORKS

An *ad hoc network* consists of a set of nodes that can talk directly to one another. For instance, WiFi devices can be configured to operate in that mode instead of the prevalent infrastructure mode where all communication takes place via an access point.

For instance, one can imagine cell phones communicating directly instead of going through a cellular base station, or wireless sensor nodes relaying each other's information until it can reach an Internet gateway. Currently, there are very few ad hoc networks other than military networks of tanks on a battlefield and mesh networks that interconnect a set of WiFi access points with wireless links, a technology that looks promising but has met only moderate commercial success.

The routing through an ad hoc network is challenging, especially when the nodes are moving. One can imagine two types of routing algorithm: proactive and reactive. A proactive algorithm calculates paths in the background to have them available when required. A reactive algorithm calculates paths only on demand, when needed. The tradeoff is between the excess traffic due to path computation messages, the delay to find a good path, and the quality of the path. Hundreds of papers compare many variations of protocols. We limit our discussion to a few examples.

5.5.1   AODV

The *AODV* (Ad Hoc On Demand Distance Vector, see [81]) routing protocol is on-demand. Essentially, if node S wants to find a path to node D, it broadcasts a route request to its neighbors: "hello, I am looking for a path to D." If one neighbor knows a path to D, it replies to node S. Otherwise, the neighbors forward the request. Eventually, replies come back with an indication of the number of hops to the destination.

The messages contain sequence numbers so that the nodes can use only the most recent information.

5.5.2   OLSR

The *Optimized Link State Routing* Protocol (OLSR, see [27]) is an adaptation of a link state protocol. The idea is that link safe messages are forwarded by a subset of the nodes instead of being flooded. Thus, OLSR is a proactive algorithm quite similar to a standard link state algorithm.

5.5.3   ANT ROUTING

Ant routing algorithms are inspired by the way ants find their way to a food source. The ants deposit some pheromone as they travel along the trail from the colony to the food and back. The pheromone evaporates progressively. Ants tend to favor trails along which the scent of pheromone is stronger. Consequently, a trail with a shorter round-trip time tends to have a stronger scent and to be selected with a higher probability by subsequent ants. Some routing algorithms use similar mechanisms, and they are called ant routing algorithms.

5.5.4   GEOGRAPHIC ROUTING

Geographic routing, as the name indicates, is based on the location of the nodes. There are many variations of such schemes. The basic idea is as follows. Say that each node knows its location and that of its neighbors. When it has to send a message to a given destination, a node selects the neighbor closest (in terms of physical distance) to that destination and forwards the message to it. There are many situations where this routing may end up in a dead end. In such a case, the node at the end of the dead end can initiate a backtracking phase to attempt to discover an alternate path.

5.5.5   BACKPRESSURE ROUTING

Backpressure routing is a form of dynamic routing that automatically adapts to changing link characteristics. We explore that mechanism in the chapter on models.

5.6   SUMMARY

Routing is the selection of the path that packets follow in the network. We explained the following ideas:

- For scalability and flexibility, the Internet uses a two-level routing scheme: nodes are grouped into Autonomous Systems (domains), and the routing is decomposed into inter-domain and intra-domain routing.

- Inter-domain routing uses a path vector protocol (BGP). This protocol enables us to adapt to inter-domain agreements of peering or transit. BGP may fail to converge if the policies of the different domains are not consistent.

- Intra-domain routing uses a distance vector protocol (based on the Bellman-Ford algorithm) or a link state protocol (based on Dijkstra's algorithm).

- The shortest path algorithms extend directly to anycast routing.

- For multicast, the shortest (Steiner) tree is not the tree of shortest paths and its computation is hard. One approach to make multicast reliable is to use a packet erasure code. Network coding can in principle increase the throughput of a multicast tree.

- Routing in ad hoc networks is challenging as the topology of the network and its link characteristic fluctuate. We explained the main ideas behind AODV, OLSR, ant routing, and geographic routing.

5.7   PROBLEMS

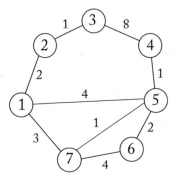

Figure 5.17: Figure for routing Problem 1.

P5.1   (a) Run Dijkstra's algorithm on the following network to determine the routing table for node 3.

(b) Repeat (a) using Bellman-Ford algorithm.

P5.2   Consider the network configuration shown in Figure 5.18. Assume that each link has the same cost.

(a) Run Bellman-Ford algorithm on this network to compute the routing table for the node A. Show A's distances to all other nodes at each step.

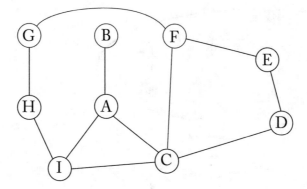

Figure 5.18: Figure for routing Problem 2.

(b) Suppose the link A-B goes down. As a result, A advertises a distance of infinity to B. Describe in detail a scenario where C takes a long time to learn that B is no longer reachable.

P5.3 Consider the network shown below and assume the following:

- The network addresses of nodes are given by <AS>.<Network>.0.<node>, e.g., node A has the address AS1.E1.0.A,
- The bridge IDs satisfy B1 < B2 < B3.
- H is not connected to AS2.E5 for part (a),
- The BGP Speakers use the least-next-hop-cost policy for routing (i.e., among alternative paths to the destination AS, choose the one that has the least cost on the first hop), and
- The network topology shown has been stable for a long enough time to allow all the routing algorithms to converge and all the bridges to learn where to forward each packet.

(a) What route, specified as a series of bridges and routers, would be used if G wanted to send a packet to A?

(b) Now, if node H was added to AS2.E5, and D tried to send a packet to it as soon as H was added, what would happen? Specifically, will the packet reach its destination and which links and/or networks would the packet be sent on?

(c) Starting from the network in (b), suppose AS2.R2 goes down. Outline in brief the routing changes that would occur as a consequence of this failure. [Hint: Think about how the change affects packets sent from AS1 to AS2, and packets sent from AS2 to AS1.]

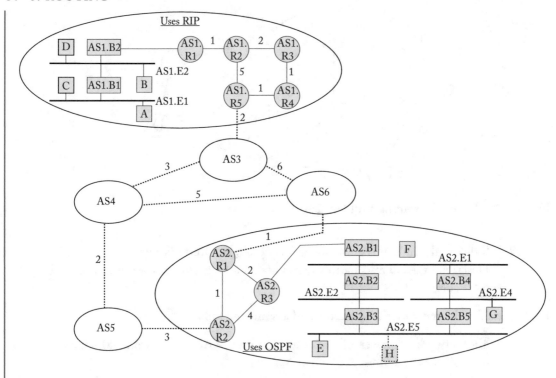

Figure 5.19: Figure for routing Problem 3.

P5.4 Consider a wireless network with nodes X and Y exchanging packets via an access point Z. For simplicity, we assume that there are no link-layer acknowledgments. Suppose that X sends packets to Y at rate 2R packets/second and Y sends packets to X at rate R packets/second; all the packets are of the maximum size allowed. The access point uses network coding. That is, whenever it can, it sends the "exclusive or" of a packet from X and a packet from Y instead of sending the two packets separately.

(a) What is the total rate of packet transmissions by the three nodes without network coding?

(b) What is the total rate of packet transmissions by the three nodes with network coding?

P5.5 Consider a set of switches as shown in Figure 5.20. Assume the following:

- The propagation time between the source and the first switch is equal to the propagation time between any pair of adjacent switches/node (denote it by T).

- Node A wants to transmit U bytes of data to node E. Node A can transmit the data either by one big message or several packets. Either message or packet must have a header of H bytes.

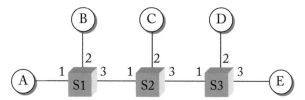

Figure 5.20: Figure for routing Problem 5.

- The transmission speed of each link is equal to S bits/second.
- Processing times in the switches are negligible.

(a) Assuming that datagram forwarding is used, provide the forwarding table for switch S1 (assuming that all routes are known).

(b) Let's define the end-to-end delay as the delay from the time A transmits the first bit until the last bit is received by E. Compute the end-to-end delay if the data is sent as a single message. Compute the end-to-end delay if the data is sent using 3 packets, each having equal payload. Explain the reasons for the difference.

(c) Compute the delay from the beginning until the last bit is received by E if the 3 packets are transmitted using a virtual circuit mode. Assume that the transmission times for the signaling messages are negligible as compared to the propagation times.

5.8   REFERENCES

Peering and transit agreements are explained in [78]. Dijkstra's algorithm was published in [29]. The Bellman-Ford algorithm is analyzed in [15]. The QoS routing problems are studied in [85]. The oscillations of BGP are discussed in [37]. BGP is described in [89]. Network coding was introduced in [8] that proved the basic multicasting result. The wireless example is from [53]. Packet erasure codes are studied in [93]. AODV is explained in [81] and OLSR in [27]. For ant routing, see [28]. For a survey of geographic routing, see [97].

CHAPTER 6

Internetworking

The Internet is a collection of networks. The key idea is to interconnect networks that possibly use different technologies, such as wireless, wired, optical, and whatever.

In this chapter, we explain how Internet interconnects Ethernet networks.

6.1 OBJECTIVE

The Internet protocols provide a systematic way for interconnecting networks. The general goal is shown in Figure 6.1. The top part of the figure shows a collection of networks, each with its own addressing and switching scheme, interconnected via routers.

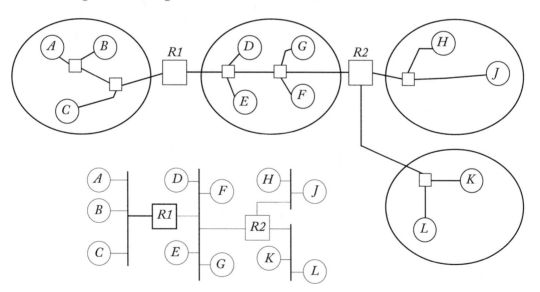

Figure 6.1: Interconnected networks.

The bottom of Figure 6.1 shows the networks as seen by the routers. The key point is that the routers ignore the details of the actual networks. Two features are essential:

- Connectivity: All the devices on any given network can send packets to one another. To do this, a device encapsulates a packet with the appropriate format for the network and the suitable addresses. The network takes care of delivering the packet.

- Broadcast-Capability: Each network is broadcast-capable. For instance, router $R1$ in Figure 6.1 can send a broadcast packet to the left-most network. That packet will reach the devices A, B, C in that network.

Examples of networks that can be interconnected with IP include Ethernet, WiFi, Cable networks, and many more. We focus on Ethernet in this chapter, but it should be clear how to adapt the discussion to other networks.

Figure 6.2 shows one Ethernet network attached to a router. The Ethernet network has the two features needed for interconnection: connectivity and broadcast-capability. The bottom of the figure shows that the router sees the Ethernet network as a link that connects it directly to all the hosts on the Ethernet.

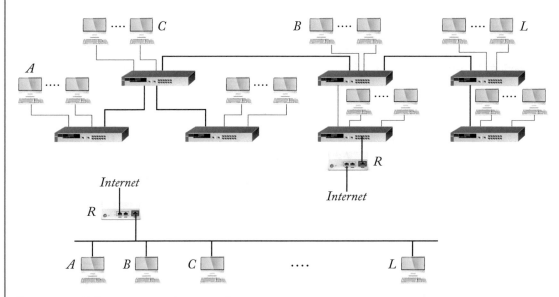

Figure 6.2: Ethernet network as seen by router.

6.2   BASIC COMPONENTS: SUBNET, GATEWAY, ARP

We illustrate the interconnection of Ethernet networks with the setup of Figure 6.3. Figure 6.3 shows two Ethernet networks: a network with devices that have MAC addresses $e1$, $e2$, and $e4$ and another network with devices that have MAC addresses $e3$ and $e5$. The two networks are attached via routers $R1$ and $R2$. A typical situation is when the two Ethernet networks belong to the same organization. For instance, these two networks could correspond to different floors in Cory Hall, on the U.C. Berkeley campus. The networks are also attached to the rest of the Internet via a port of R2. The rest of the Internet includes some other Ethernet networks of the Berkeley campus in our example.

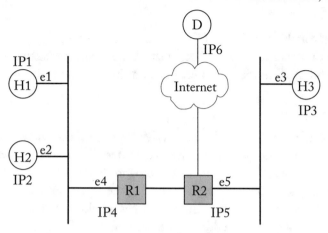

Figure 6.3: Interconnected Ethernet networks.

6.2.1   ADDRESSES AND SUBNETS

In addition to a MAC address, each device is given an IP address, as shown in the figure. This address is based on the location. Moreover, the addresses are organized into *subnets*. For instance, in Figure 6.3 the Ethernet network on the left is one subnet and the one on the right is another subnet. The IP addresses of the devices $\{e1, e2, e4\}$ on the first subnet have a different prefix (initial bit string) than the devices $\{e3, e5\}$ on the second subnet.

The *subnet address* specifies the leading bits in the IP addresses that are common to all the device interfaces attached to that Ethernet network. For instance, say that the subnet addresses of the Ethernet networks are the first 24 bits. This means that two IP addresses IPi and IPj of interfaces attached to the same Ethernet network have the same first 24 bits. If IPi and IPj are not on the same Ethernet network, then their first 24 bits are not the same. In the figure, the first 24 bits of IP1, which we designate by IP1/24, are the same as IP2/24 but not the same as IP3/24.

6.2.2   GATEWAY

The *gateway router* of an Ethernet network is the router that connects that Ethernet network to the rest of the Internet. For instance, the gateway router of the Ethernet network on the left (with devices H1, H2) is R1. Host H1 knows the IP address of the interface of R1 on its Ethernet network. The gateway router for the network on the right is R2.

6.2.3 DNS SERVER

The devices know the address of a local DNS server. For instance, if the Ethernet networks are in Cory Hall, the local Ethernet server is that of the domain eecs.berkeley.edu. We discussed the basic concepts behind DNS in the earlier chapters on the Internet architecture and principles.

6.2.4 ARP

The *Address Resolution Protocol*, or ARP, enables devices to find the MAC address of interfaces on the *same* Ethernet that corresponds to a given IP address. Here is how it works. Say that host H1 wants to find the MAC address that corresponds to the IP address IP2. Host H1 sends a special broadcast ARP packet on the Ethernet. That is, the destination address is the "Broadcast MAC Address" represented by 48 ones. The packet has a field that specifies that it is an "ARP request," and it also specifies the IP address IP2. Thus, the ARP request has the form [all | e1 | who is IP2?], where the addresses mean "to all devices, from e1." An Ethernet switch that receives a broadcast packet repeats the packet on all its output ports. When a device gets the ARP request packet, it compares its own IP address with the address IP2. If the addresses are not identical, the device ignores the request. If IP2 is the IP address of the device, then it answers the request with an Ethernet packet [e1 | e2 | I am IP2]. The source address of that reply tells H1 that e2 is the MAC address that corresponds to IP2. Note that a router does not forward a packet with a broadcast Ethernet address.

6.2.5 CONFIGURATION

Summarizing, to attach an Ethernet network on the Internet, each device must have an IP address, know the subnet address, know the IP address of the gateway router and the IP address of the DNS server. In addition, each device must implement ARP.

6.3 EXAMPLES

We explore how a device sends a packet to another device on the subnets we discussed for the interconnected networks shown in Figure 6.3. We examine two separate cases: first, when the devices are on the same subnet; second, when they are on different subnets.

6.3.1 SAME SUBNET

We first examine how device H1 sends a packet to H2 assuming that H1 knows IP2 (see Figure 6.4). Since IP1/24 = IP2/24, H1 knows that IP2 is on the same Ethernet as H1. Accordingly, H1 needs to send the packet as an Ethernet packet directly to H2. To do this, H1 needs the MAC address of IP2. It finds that address using ARP, as we explained above. Once it finds out e2, H1 forms the packet [e2 | e1 | IP1 | IP2 | X] where e2 is the destination MAC address, e1 is the source MAC address, IP1 is the source IP address, IP2 is the destination IP address, and

X is the rest of the packet. Note that the IP packet [IP1 | IP2 | X] from H1–H2 is *encapsulated* in the Ethernet packet with the MAC addresses [e2 | e1].

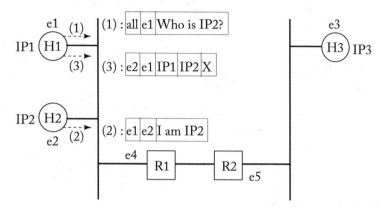

Figure 6.4: Sending on the same subnet.

6.3.2 DIFFERENT SUBNETS

We explain how H1 sends a packet to H3, assuming that H1 knows IP3. (See Figure 6.5.) Since IP1/24 ≠ IP3/24, H1 knows that H3 is not on the same Ethernet as H1. H1 must then send the packet first to the gateway. Using the IP address of the gateway R1, say IP4, H1 uses ARP to find the MAC address e4. H1 then sends the packet [e4 | e1 | IP1 | IP3 | X] to R1. Note that this is the IP packet [IP1 | IP3 | X] from H1–H3 that is encapsulated in an Ethernet packet from H1–R1. When it gets that packet, R1 decapsulates it to recover [IP1 | IP3 | X], and it consults its routing table to find the output port that corresponds to the destination IP3. R1 then sends that packet out on that port to R2. On that link, the packet is encapsulated in the appropriate format. (The figure shows a specific link header on that link.) When it gets the packet [IP1 | IP3 | X], router R2 consults its routing table and finds that IP3 is on the same Ethernet as its interface e5. Using ARP, R2 finds the MAC address e3 that corresponds to IP3, and it sends the Ethernet packet [e3 | e5 | IP1 | IP3 | X]. Note that the IP packet [IP1 | IP3 | X] is not modified across the different links but that its encapsulation changes.

6.3.3 FINDING IP ADDRESSES

So far, we have assumed that H1 knows IP2 and IP3. What if H1 only knows the name of H2, but not its IP address IP2? In that case, H1 uses DNS to find IP2. Recall that H1 knows the IP address of the DNS server. Using the same procedure as the one we just explained, H1 can send a DNS request to discover IP2, and similarly for IP3.

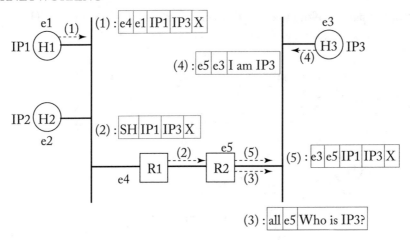

Figure 6.5: Sending on a different subnet.

6.3.4   FRAGMENTATION

In general, an IP packet might need to be *fragmented*. Say that [IP1 | IP3 | X] has 1,500 bytes, which is allowed by Ethernet networks. Assume that the link between routers R1 and R2 can only transmit 500 Byte packets (as could be the case for some wireless links). In that case, the network protocol IP fragments the packet and uses the header of the packet to specify the packet identification and where the fragments fit in the packet. It is up to the destination H3 to put the fragments back together.

The details of how the header specifies the packet and the fragment positions are not essential for us. Let us simply note that each IP packet contains, in addition to the addresses, some control information such as that needed for fragmentation and reassembly by the destination. All this information is in the *IP header* which is protected by an error detection code.

6.4   DHCP

When you need to attach your laptop to some Ethernet or a WiFi network, that laptop needs an IP address consistent with that network. When you attach the laptop to another network, it needs another IP address. Instead of reserving a permanent IP address for your laptop, the network maintains a pool of addresses that it assigns when needed. The protocol is called *Dynamic Host Configuration Protocol* (DHCP), and it works as follows.

When one attaches the laptop to the network, the laptop sends a DHCP request asking for an IP address. The network has a DHCP server commonly implemented by a router in the network that maintains the list of available IP addresses. The DHCP server then allocates the address to the laptop. Periodically, the laptop sends a request to renew the lease on the IP address.

If it fails to do so, as happens when you turn off your laptop or you move it somewhere else, the server puts the address back in the pool of available addresses.

DHCP is also commonly used by ISPs to assign addresses to the devices of their customers. This mechanism reduces the number of addresses that the ISP must reserve.

Note that a device that gets an IP address with the DHCP protocol has a temporary address that other devices do not know. Consequently, that device cannot be a server.

6.5   NAT

In the late 1990s, Internet engineers realized that one would run out of IP addresses in a few years. They then developed a new addressing scheme that uses 128 bits instead of 32. However, these new addresses require a new version of the Internet Protocol, which necessitates some considerable amount of configuration work in thousands of routers. As explained above, DHCP is one mechanism that reduces the number of IP addresses needed by allocating them temporarily instead of permanently. The *Network Address Translation* (NAT) is another scheme that enables us to reuse IP addresses.

Most home routers implement a NAT. With that device, the devices in the home network use a set of IP addresses, called *Private Addresses*, that are also used by many other home networks. Figure 6.6 explains how NAT works. The trick is to use the *port numbers* of the transport protocol. To be able to associate byte streams or datagrams with specific applications inside a

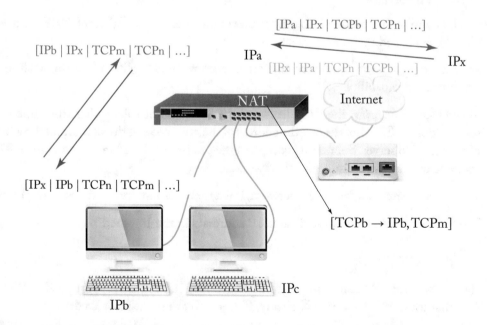

Figure 6.6: Network address translation.

computer, the transport protocol uses a port number that it specifies in the transport header of the packet. Thus, at the transport layer, a packet has the following structure:

[source IP | destination IP | source port | destination port | $\cdots$ | data].

For instance, the HTTP protocol uses port 80 whereas email uses port 25.

The NAT uses the port number as follows. The private addresses of devices inside the home are IPb, IPc. The NAT has an address IPa. Assume that IPb sends a packet [IPb | IPx | TCPm | TCPn | $\cdots$] to a device with IP address IPx. In this packet, TCPm is the source port number and TCPn the destination port number. The NAT converts this packet into [IPa | IPx | TCPb | TCPn | $\cdots$] where TCPb is chosen by the NAT device which notes that TCPb corresponds to (IPb, TCPm). When the destination with address IPx and port TCPn replies to this packet, it sends a packet [IPx | IPa | TCPn | TCPb | $\cdots$]. When the NAT gets this packet, it maps back the port number TCPb into the pair (IPb, TCPm), and it sends the packet [IPx | IPb | TCPn | TCPm | $\cdots$] to device IPb.

Note that this scheme works only when the packet is initiated inside the home. It is not possible to send a request directly to a device such as IPb, only to reply to requests from that device. Some web servers exist that maintain a connection initiated by a home device so that it is reachable from outside.

6.6   SUMMARY

The main function of the Internet is to interconnect networks such as Ethernet, WiFi, and cable networks. We explained the main mechanisms.

- Internet interconnects "local area networks," each with its specific addressing technology and broadcast capability.

- When using *subnetting*, the IP addresses of the hosts on one subnet share the same prefix subnet address. Each host knows the address of the router to exit the subnet and the address of a local DNS server. Subnetting compares the prefixes of the addresses and uses ARP to translate the IP address into a local network address.

- When a computer attaches to the network, it may get a temporary IP address using DHCP.

- The NAT enables the use of replicated IP addresses in the Internet.

6.7   PROBLEMS

P6.1   (a) How many IP addresses need to be leased from an ISP to support a DHCP server that uses NAT to service N clients, if every client uses at most P ports?

(b) If M unique clients request an IP address every day from the above mentioned DHCP server, what is the maximum lease time allowable to prevent new clients from being denied access, assuming that requests are uniformly spaced throughout the day, and that the addressing scheme used supports a max of N clients?

P6.2 Below is the DNS record for a fictitious corporation, OK Computer:

Table 6.1: Table for internetworking Problem 2

Name	Type	Value	TTL (Seconds)
okcomputer.com	A	164.32.15.98	86,400 (1 day)
okcomputer.com	NS	thom.yorke.net	86,400
okcomputer.com	NS	karma.okcomputer.com	86,400
okcomputer.com	MX	android.okcomputer.com	60
lucky.okcomputer.com	A	164.32.12.8	86,400
www.okcomputer.com	CNAME	lucky.okcomputer.com	86,400
android.okcomputer.com	A	164.32.15.99	86,400

(a) If you type http://www.okcomputer.com into your web browser, to which IP address will your web browser connect?

(b) If you send an e-mail to thom@okcomputer.com, to which IP address will the message get delivered?

(c) The TTL field refers to the maximum amount of time a DNS server can cache the record. Give a rationale for why most of the TTLs were chosen to be 86,400 s (1 day) instead of a shorter or a longer time, and why the MX record was chosen to have a 60-second TTL?

6.8 REFERENCES

The Address Resolution Protocol is described in [82], subnetting in [73], DHCP in [30], and NAT in [102].

CHAPTER 7

Transport

The transport layer supervises the end-to-end delivery across the Internet between a process in a source device and a process in a destination device. The transport protocol of Internet implements two transport services: a connection-oriented reliable byte stream delivery service and a connectionless datagram delivery service. This chapter explains the main operations of this layer.

7.1 TRANSPORT SERVICES

The network layer of the Internet (the Internet Protocol, IP) provides a basic service of packet delivery from one host to another host. This delivery is not reliable. The transport layer adds a few important capabilities such as multiplexing, error control, congestion control, and flow control, as we explain in this chapter.

Figure 7.1 shows the protocol layers in three different Internet hosts. The transport layer defines *ports* that distinguish information flows. These are logical ports, not to be confused with the physical ports of switches and routers. The host on the left has two ports $p1$ and $p2$ to which application processes are attached. Port $p2$ in that host is communicating with port $p1$ in the middle host. The protocol HTTP is attached to port $p1$ in the middle host.

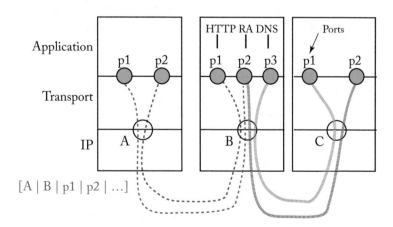

Figure 7.1: The transport layer supervises the delivery of information between *ports* in different hosts.

The ports are identified by a number from 1–65,535 and are of three types: well-known ports (1–1,023) that correspond to fixed processes, registered ports (1,024–49,451) that have been registered by companies for specific applications, and the dynamic and/or private ports that can be assigned dynamically. For instance, the basic email protocol SMTP is attached to port 25, HTTP to port 80, the real time streaming protocol RTSP to port 540, the game Quake Wars to port 7,133, Pokemon Netbattle to port 30,000, traceroute to 33,434, and so on.

Thus, at the transport layer, information is delivered from a source port in a host with a source IP address to a destination port in a host with some other IP address.

The transport layer implements two protocols between two hosts: UDP (the *User Datagram Protocol*) and TCP (the *Transmission Control Protocol*) with the following characteristics:

- UDP delivers individual packets. The delivery is not reliable.

- TCP delivers a byte stream. The byte stream is reliable: the two hosts arrange for retransmission of packets that fail to arrive correctly. Moreover, the source regulates the delivery to control congestion in routers (congestion control) and in the destination device (flow control).

Summarizing, an information flow is identified by the following set of parameters: (source IP address, source transport port, destination IP address, destination transport port, protocol), where the protocol is UDP or TCP. The port numbers enable us to *multiplex* the packets of different applications that run on a given host.

7.2   TRANSPORT HEADER

Figure 7.2 shows the header of every UDP packet. The header specifies the port numbers and the length of the UDP packet (including UDP header and payload, but not the IP header). The header has an optional UDP checksum calculated over the UDP packet.

0	16	31
Source Port	Destination Port	
UDP Length	UDP Checksum	
Payload (variable)		

Figure 7.2: The header of UDP packets contains the control information.

Each TCP packet has a header shown in Figure 7.3. In this header, the sequence number, acknowledgment, and advertised window are used by the Go Back N protocol explained later in this chapter. The "Flags" are as follows:

- SYN, FIN: establishing/terminating a TCP connection;

Source Port	Destination Port
Sequence Number			
Acknowledgment			
HdrLen	Reserved	Flags	Advertised Window
Checksum			Urgent Pointer
Options (variable)			
Payload (variable)			

Figure 7.3: The header of TCP packets contains the control information.

- ACK: set when Acknowledgment field is valid;

- URG: urgent data; Urgent Pointer says where non-urgent data starts;

- PUSH: don't wait to fill segment; and

- RESET: abort connection.

The precise functions of the main fields are described below. The UDP header is a stripped-down version of a TCP header since its main function is multiplexing through port numbers.

7.3 TCP STATES

A TCP connection goes through the following phases: setup, exchange data, close. (See Figure 7.4.)

The setup goes as follows. The client sets up a connection with a server in three steps:

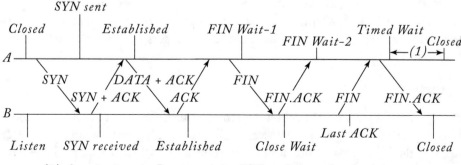

(1) A waits in case B retransmits FIN and A mush ack again

Figure 7.4: The phases and states of a TCP connection.

- The client sends a SYN packet (a TCP packet with the SYN flag set identifying it as a SYNchronization packet). The SYN packet specifies a random number X.

- The server responds to the SYN with a SYN.ACK (i.e., by setting both SYN and ACK flags) that specifies a random number Y and an ACK with sequence number $X + 1$.[1]

- The client sends the first data packet with sequence number $X + 1$ and an ACK with sequence number $Y + 1$.

The hosts then exchange data and acknowledge each correct data packet, either in a data packet or in a packet that contains only an ACK. When host A has sent all its data, it closes its connection to the other host, host B, by sending a FIN; A then waits for a FIN.ack. Eventually, host B also closes its connection by sending a FIN and waiting for a FIN.ack. After sending its FIN.ack, host A waits to make sure that host B does not send a FIN again that host A would then acknowledge. This last step is useful in case the FIN.ack from A gets lost.

7.4 ERROR CONTROL

When using TCP, the hosts control errors by retransmitting packets that are not acknowledged before a timeout. We first describe a simplistic scheme called Stop-and-Wait and then explain the Go Back N mechanism of TCP.

7.4.1 STOP-AND-WAIT

The simplest scheme of retransmission after timeout is *stop-and-wait*. When using that scheme, the source sends a packet, waits for up to T seconds for an acknowledgment, retransmits the packet if the acknowledgment fails to arrive, and moves on to the next packet when it gets the acknowledgment. Even if there are no errors, the source can send only one packet every T s, and T has to be as large as a round-trip time across the network. A typical round trip time in the Internet is about 40 ms, If the packets have 1,500 bytes, this corresponds to a rate equal to $1,500 \times 8/0.04 = 300$ Kbps. If the link rates are larger, this throughput can be increased by allowing the source to send more than one packet before it gets the first acknowledgment. That is precisely what Go Back N does.

7.4.2 GO BACK N

The basic version of TCP uses the following scheme called *Go Back N* and illustrated in Figure 7.5. The top part of the figure shows a string of packets with sequence numbers $1, 2, \ldots$ that the source (on the left) sends to the destination (on the right). In that part of the figure, the destination has received packets $1, 2, 3, 4, 5, 7, 8, 9$. It is missing packet 6.

---

[1]For simplicity, we talk about the packet sequence numbers. However, TCP actually implements the sequence numbers as the ones associated with the bytes transported.

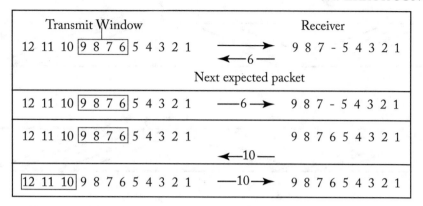

Figure 7.5: Go back N protocol.

The scheme specifies that, at any time, the source can have sent up to N packets that have not yet been acknowledged ($N = 4$ in the figure). When it gets a packet, the destination sends an ACK with the sequence number of the next packet it expects to receive, *in order*. Thus, if the receiver gets the packets 1, 2, 3, 4, 5, 7, 8, 9, it sends the ACKs 2, 3, 4, 5, 6, 6, 6, 6 as it receives those packets. If an ACK for a packet fails to arrive after some time, the source retransmits that packet and possibly the subsequent $N - 1$ packets.

The source slides its window of size N so that it starts with the last packet that has not been acknowledged in sequence. This window, referred to as Transmit Window in the figure, specifies the packets that the source can transmit. Assume that the source retransmits packet 6 and that this packet arrives successfully at the receiver. The destination then sends back an acknowledgment with sequence number 10. When it gets this ACK, the source moves its transmission window to [10, 11, 12, 13] and sends packet 10. Note that the source may retransmit packets 7, 8, 9 right after packet 6 before the ACK 10 arrives.

If there are no errors, Go Back N sends N packets every round trip time (assuming that the N transmission times take less than one round trip time). Thus, the throughput of this protocol is N times larger than that of Stop-and-Wait.

7.4.3 SELECTIVE ACKNOWLEDGMENTS

Go Back N may retransmit packets unnecessarily, as shown in Figure 7.6. The figure shows the packets and acknowledgments transmitted by the source and receiver, respectively, along with the corresponding packet or acknowledgment numbers. The top of the figure shows Go Back N with a window of size 4 when the second packet gets lost. After some time, the sender retransmits packets 2, 3, 4 and transmits packet 5.

The destination had already received packets 3 and 4, so that these retransmissions are unnecessary; they add delay and congestion in the network. To prevent such unnecessary re-

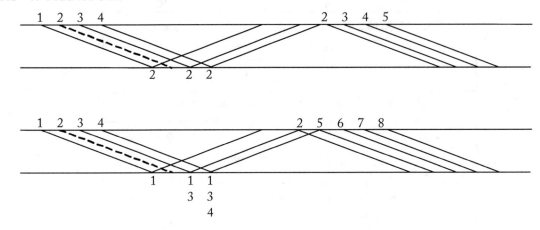

Figure 7.6: Comparing Go Back N (top) and selective ACKs (bottom).

transmissions, a version of TCP uses *selective acknowledgments*, as illustrated in the bottom part of the figure. When using this scheme, the receiver sends a positive ACK for all the packets it has received correctly. Thus, when it gets packet 3, the receiver sends an ACK that indicates it received packets 1 and 3. When the sender gets this ACK, it retransmits packet 2, and so on. The figure shows that using selective acknowledgments increases the throughput of the connection.

A special field in the SYN packet indicates if the sender accepts selective ACKs. The receiver then indicates in the SYN.ACK if it will use selective ACKs. In that case, fields in the TCP header of ACK packets indicate up to four blocks of contiguous bytes that the receiver got correctly. For instance, the fields could contain the information [3; 1001, 3000; 5001, 8000; 9001, 12000] which would indicate that the receiver got all the bytes from 1,001 to 3,000, also from 5,001–8,000, and from 9,001–12,000. The length field that contains the value 3 indicates that the remaining fields specify three blocks of contiguous bytes. See [69] for details.

7.4.4 TIMERS

How long should the source wait for an acknowledgment? Since the round trip time varies greatly from one connection to another, the timeout value must be adapted to the typical round trip time of the connection. To do this, the source measures the round trip times $T_n, n = 1, 2, \ldots$ for the successive packets. That is, T_n is the time between the transmission of the n-th packet and the reception of the corresponding acknowledgment. (The source ignores the packets that are not acknowledged in this estimation.) The source then calculates the average value A_n of the round-trip times $\{T_1, \ldots, T_n\}$ and the average deviation D_n around that mean value. It then sets the timeout value to $A_n + 4D_n$. The justification is that it is unlikely that a round trip time exceeds $A_n + 4D_n$ if the acknowledgment arrives at all.

The source calculates A_n and D_n recursively as exponential averages defined as follows ($b < 1$ is a parameter of the algorithm):

$$A_{n+1} = (1 - b)A_n + bT_{n+1} \text{ and } D_{n+1} = (1 - b)D_n + b|T_{n+1} - A_{n+1}|, n \geq 1$$

with $A_1 = T_1$ and $D_1 = T_1$.

Figure 7.7 shows the exponential averaging of times T_n for different values of the parameter b. As the graph shows, when b is small, A_n calculates the average of the T_n over a longer period of time and is less sensitive to the recent values of those times.

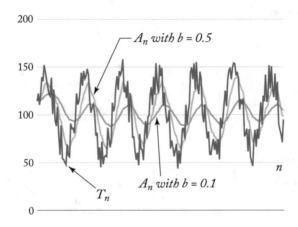

Figure 7.7: Exponential averaging.

7.5 CONGESTION CONTROL

In the Internet, multiple flows share links. The devices do not know the network topology, the bandwidth of the links, nor the number of flows that share links. The challenge is to design a congestion control scheme that enables the different flows to share the links in a reasonably fair way.

7.5.1 AIMD

The TCP congestion algorithm is AIMD (additive increase—multiplicative decrease). This scheme attempts to share the links fairly among the connections that use them.

Consider two devices A and B shown in Figure 7.8 that send flows with rates x and y that share a link with bandwidth C. The sources increase x and y additively as long as they receive acknowledgments. When they miss acknowledgments, they suspect that a router had to drop packets, which happens here soon after $x + y$ exceeds C. The sources then divide their rate by 2 and resume increasing them. Following this procedure, say that the initial pair of rates (x, y)

corresponds to the point 1 in the right part of the figure. Since the sum of the rates is less than C, the link buffer does not accumulate a backlog, the sources get the acknowledgments, and they increase their rate linearly over time, at the same rate (ideally). Accordingly, the pair of rates follows the line segment from point 1–2. When the sum of the rates exceeds C, the buffer arrival rate exceeds the departure rate and the buffer starts accumulating packets. Eventually, the buffer has to drop arriving packets when it runs out of space to store them. A while later, the sources notice missing acknowledgments and they divide their rate by a factor 2, so that the pair of rates jumps to point 3, at least if we assume that the sources divide their rate at the same time. The process continues in that way, and one sees that the pair of rates eventually approaches the set S where the rates are equal.

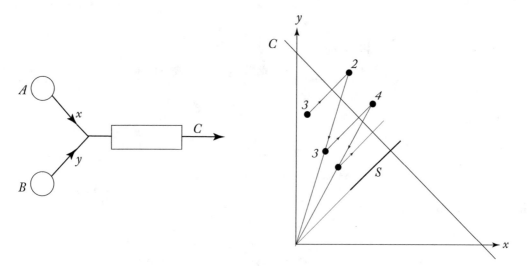

Figure 7.8: Additive increase, multiplicative decrease.

The scheme works for an arbitrary number of flows. The sources do not need to know the number of flows or the rate of the link they share. You can verify that a scheme that would increase the rates multiplicatively or decrease them additively would not converge to equal rates. You can also check that if the rates do not increase at the same pace, the limiting rates are not equal.

7.5.2   REFINEMENTS: FAST RETRANSMIT AND FAST RECOVERY

Assume that the destination gets the packets 11, 12, 13, 15, 16, 17, 18. It then sends the ACKs 12, 13, 14, 14, 14, 14, 14. As it gets duplicate ACKs, the source realizes that packet 14 has failed to arrive. The *fast retransmit* scheme of TCP starts retransmitting a packet after three duplicate ACKs, thus avoiding having to wait for a timeout.

The *fast recovery* scheme of TCP is designed to avoid having to wait a round-trip time after having divided the window size by 2. To see how this scheme works, first note how the "normal" (i.e., without fast recovery) scheme works (see the top part of Figure 7.9). Assume that the receiver gets the packets 99, 100, 102, 103, . . . , 132 and sends the ACKs 100, 101, 101, 101, . . . , 101. Assume also that the window size is 32 when the sender sends packet 132. When the source gets the third duplicate ACK with sequence number 101, it reduces its window to $32/2 = 16$ and starts retransmitting 101, 102, 103, . . . , 117. Note that the source has to wait one round-trip time after it sends the copy of 101 to get the ACK 133 and to be able to transmit 133, 134, . . .

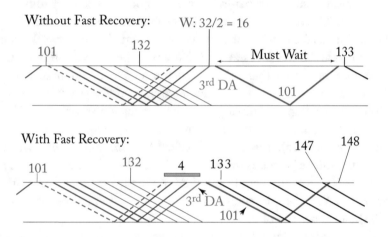

Figure 7.9: TCP without (top) and with fast recovery.

A better scheme, called *fast recovery* is shown in the bottom part of Figure 7.9. With this scheme, the source sends packets 101, 133, 134, . . . , 147 by the time it gets the ACK of the copy of 101 (with sequence number 133). In that way, there are exactly 15 unacknowledged packets (133, 134, . . . , 147). With the new window of $32/2 = 16$, the source is allowed to transmit Packet 148 immediately, and the normal window operation can resume with the window size of 16. Note that the idle waiting time is eliminated with this scheme.

To keep track of what to do, the source modifies its window as follows. When it gets the third duplicate ACK 101, the sender sends a copy of 101 and changes its window size from $W = 32$ to $W/2 + 3 = 32/2 + 3 = 19$. It then increases its window by one whenever it gets another duplicate ACK of 101. Since the source already got the original and three duplicate ACKs from the $W = 32$ packets 100, 102, 103, . . . , 132, it will receive another $W - 4 = 28$ duplicate ACKs and will increase its window by $W - 4$ to $W/2 + 3 + (W - 4) = W + W/2 - 1 = 47$, and it can send all the packets up to 147 since the last ACK it received is 100 and the set of unacknowledged packets is then $\{101, 102, . . . , 147\}$. Once the lost packet is acknowledged, the window size is set to $W/2$, and since now there are $W/2 - 1$ outstanding packets, the source can send the next packet immediately.

7.5.3   ADJUSTING THE RATE

As long as no packet gets lost, the sliding window protocol sends N packets every round trip time (the time to get an ACK after sending a packet). TCP adjusts the rate by adjusting the window size N. The basic idea is to approximate AIMD. Whenever it gets an ACK, the sender replaces N by $N + 1/N$. In a round-trip time, since it gets N ACKs, the sender ends up adding approximately $1/N$ about N times during that round-trip time, thus increasing the window size by 1 packet. Thus, the source increases its rate by about 1 packet per round trip time every round trip time. This increase in the window size enables the sources to take full advantage of the available bandwidth as the number of connections changes in the network. When the source misses an ACK, it divides its window size by 2. Note that the connections with a shorter round trip time increase their rate faster than others, which results in an unfair advantage.

This scheme might take a long time to increase the window size to the acceptable rate on a fast connection. To speed up the initial phase, the connection starts by doubling its window size every round trip time, until it either misses an ACK or the window size reaches a threshold (as discussed in the next section). This scheme is called *Slow Start*. To double the window size in a round-trip time, the source increases the window size by 1 every time it gets an ACK. Thus, if the window side was N, the source should get N ACKs in the next round-trip time and increase the window by N, to $2N$. When the source misses an ACK, after a timeout, it restarts the slow start phase with the window size of 1.

When using selective acknowledgments, the window can grow/shift without worrying about the lost packets. That is, if the window size is 150 and if the acknowledgments indicate that the receiver got bytes [0, 1199], [1250, 1999], [2050, 2999], then the source can send bytes [1200, 1249], [2000, 2049], [3000, 3049]. Note that the gap between the last byte that was not acknowledged (1200) and the last byte that can be sent (3049) is larger than the window size, which is not possible when using the Go Back N with cumulative acknowledgments. The rules for adjusting the window are the same as with cumulative acknowledgments.

7.5.4   TCP WINDOW SIZE

Figure 7.10 shows how the TCP window size changes over time. During the slow start phase (labeled SS in the figure), the window increases exponentially quickly, doubling every round-trip time. When a timeout occurs when the window size is W_0, the source remembers the value of $W_0/2$ and restarts with a window size equal to 1 and doubles the window size every round-trip time until the window size reaches $W_0/2$. After that time, the protocol enters a *congestion avoidance* phase (labeled CA). During the CA phase, the source increase the window size by one packet every round-trip time. If the source sees three duplicate ACKs, it performs a fast retransmit and fast recovery. When a timeout occurs, the source restarts a slow start phase.

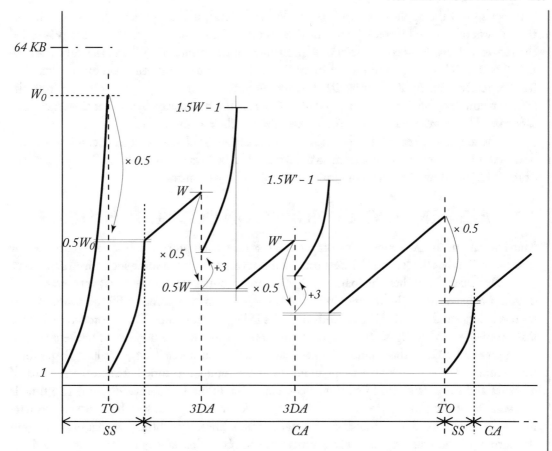

Figure 7.10: Evolution in time of the TCP window.

7.5.5 TERMINOLOGY

The various modifications of TCP received new names. The original version, with Go Back N, is called TCP Tahoe. With fast retransmit, the protocol is called TCP Reno. When fast recovery is included, the protocol is called TCP New Reno; it is the most popular implementation. When selective ACKs are used, the protocol is TCP-SACK. There are also multiple other variations that have limited implementations.

7.6 FLOW CONTROL

Congestion control is the mechanism to prevent saturating routers. Another mechanism, called *flow control*, prevents saturating the destination of an information flow and operates as follows in TCP. The end host of a TCP connection sets aside some buffer space to store the packets it receives until the application reads them. When it sends a TCP packet, that host indicates the amount of free buffer space it currently has for that TCP connection. This quantity is called the *Receiver Advertised Window* (RAW). The sender of that connection then calculates $RAW - OUT$ where OUT is the number of bytes that it has sent to the destination but for which it has not received an ACK. That is, OUT is the number of *outstanding bytes* in transit from the sender to the receiver. The quantity $RAW - OUT$ is the number of bytes that the sender can safely send to the receiver without risking overflow of the receiver buffer.

The sender then calculates the minimum of $RAW - OUT$ and its current congestion window $-OUT$ and uses that minimum to determine the packets it can transmit. This adjustment of the TCP window combines congestion control and flow control.

7.7   ALTERNATIVE CONGESTION CONTROL SCHEMES

Many variations on the congestion control or TCP have been proposed. They are motivated by the realization that the "normal" TCP protocol is not very efficient in some specialized situations.

One example is when the "bandwidth-distance" product of a connection is very large, TCP does not behave very well. The reason is that the window size required to fill the connection is then very large and the AIMD mechanism results in large fluctuations. One approach to remedy this problem is TCP-Vegas. This protocol attempts to estimate the number of packets queued in routers and adjusts the window to keep that number at a small target value. To perform this estimation, the source host keeps monitoring the round-trip time between the sending of a packet and the reception of its acknowledgment. The minimum observed round-trip time is an estimate of the propagation time. Given the link rate of the sending host, it can estimate how many packets can be in transit on the transmission lines. By subtracting that number from the current window size, the host can estimate the number of packets queued in routers and can adjust its window size to keep that number small. As you may suspect, this mechanism is subject to some potential difficulties. One of them is "congestion creep" where the minimum observed round-trip time already includes some congestion. Nevertheless, this mechanism is effective for fast long-distance connections.

Another situation where TCP benefits from some tuning is in specific networks with known links and topology, as in a data center. In such situations, the ideal window size and the typical round trip times are easy to determine. Using these values instead of the adaptive timeout values and default window sizes results in a more efficient protocol.

Some other networks benefit from modified versions of TCP. For instance, some manufacturing automation networks increase the reliability of packet transmissions by duplicating packets that travel along distinct paths and may therefore arrive out of order. As we know, TCP reacts poorly when packets arrive out of order because the destination generates duplicate acknowledgments. Thus, in such a network, it is beneficial to modify that aspect of TCP.

In wireless networks, packets get lost because of transmission errors. By design, TCP assumes that lost packets are due to congestion. Hence, TCP slows down when transmission errors corrupt packets, which exacerbates the loss in throughput. Researchers have proposed schemes to correct this problem by opening multiple TCP connections in parallel and adjusting the number of connections based on the observed throughput.

7.8   SUMMARY

The Transport Layer is a set of end-to-end protocols that supervise the delivery of packets.

- The transport layer offers two services: reliable byte stream (with TCP) and unreliable packet delivery (with UDP) between ports.

- A TCP connection goes through the following phases: open (with a three-way hand-shake: SYN, SYN.ACK, then ACK and start); data exchange (with data and acks); half-close (with FIN and FIN.ACK); then a second half-close (with FIN, timed wait, and FIN.ACK).

- The error control in TCP uses a sliding window protocol based on Go Back N with either cumulative or selective ACKs.

- Flow control uses the receiver-advertised window.

- Congestion control attempts to share the links equally among the active connections. The mechanism is based on additive increase, multiplicative decrease. The timers are adjusted based on the average of the round trip times and the average fluctuation around the average value.

- Refinements include fast retransmit after three duplicate ACKs and fast recovery by increasing the window size while waiting for the ACK of a retransmitted packet.

7.9   PROBLEMS

P7.1 (a) Suppose you and two friends named Alice and Bob share a 200 Kbps DSL connection to the Internet. You need to download a 100 MB file using FTP. Bob is also starting a 100 MB file transfer, while Alice is watching a 150 Kbps streaming video using UDP. You have the opportunity to unplug either Alice or Bob's computer at the router, but you can't unplug both. To minimize the transfer time of your file, whose computer should you unplug and why? Assume that the DSL connection is the only bottleneck link, and that your connection and Bob's connection have a similar round trip time.

(b) What if the rate of your DSL connection were 500 Kbps? Again, assuming that the DSL connection were the only bottleneck link, which computer should you unplug?

P7.2 Suppose Station A has an unlimited amount of data to transfer to Station E. Station A uses a sliding window transport protocol with a fixed window size. Thus, Station A begins

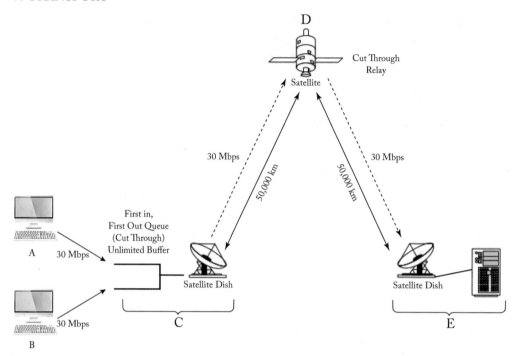

Figure 7.11: Figure for transport Problem 2.

a new packet transmission whenever the number of unacknowledged packets is less than W and any previous packet being sent from A has finished transmitting.

The size of the packets is 10,000 bits (neglect headers). So, for example, if $W > 2$, station A would start sending packet 1 at time $t = 0$, and then would send packet 2 as soon as packet 1 finished transmission, at time $t = 0.33$ ms. Assume that the speed of light is 3×10^8 m/s.

(a) Suppose station B is silent, and that there is no congestion along the acknowledgment path from C–A. (The only delay acknowledgments face is the propagation delay to and from the satellite.) Plot the average throughput as a function of window size W. What is the minimum window size that A should choose to achieve a throughput of 30 Mbps? Call this value W^*. With this choice of window size, what is the average packet delay (time from leaving A to arriving at E)?

(b) Suppose now that station B also has an unlimited amount of data to send to E, and that station B and station A both use the window size W^*. What throughput would A and B get for their flows? How much average delay do packets of both flows incur?

(c) What average throughput and delays would A and B get for their flows if A and B both used window size $0.5W^*$? What would be the average throughput and delay for each flow if A used a window size of W^* and B used a window size of $0.5W^*$?

P7.3 As shown in the Figure 7.12, flows 1 and 2 share a link with capacity $C = 120$ Kbps. There is no other bottleneck. The round trip time of flow 1 is 0.1s and that of flow 2 is 0.2s. Let x_1 and x_2 denote the rates obtained by the two flows, respectively. The hosts use AIMD to regulate their flows. That is, as long as $x_1 + x_2 < C$, the rates increase linearly over time: the window of a flow increases by one packet every round trip time. Rates are estimated as the window size divided by the round-trip time. Assume that as soon as $x_1 + x_2 > C$, the hosts divide their rates x_1 and x_2 by the factor $\alpha = 1.1$.

(a) Draw the evolution of the vector (x_1, x_2) over time.

(b) What is the approximate limiting value for the vector?

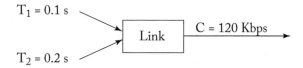

Figure 7.12: Figure for transport Problem 3.

P7.4 Consider a TCP connection between a client C and a server S.

(a) Sketch a diagram of the window size of the server S as a function of time.

(b) Using the diagram, argue that the time to transmit N packets from S to C is approximately equal to $a + bN$ for large N.

(c) Explain the key factors that determine the value of b in that expression for an Internet connection between two computers that are directly attached by an Ethernet link.

(d) Repeat question c when one of the computers is at UCB and the other one at MIT.

P7.5 Source S uses the sliding window protocol for sending data to destination D as shown in the Figure 7.13. Acknowledgments (Acks) flow from D to S over an unspecified network as shown. Assume the following: Propagation Speed $= 2 \times 10^8$ m/s, Packet Size $= 1$ KB, Ack Delay (from the time an Ack begins to be sent out from D until it is fully received at S) $= 2.2$ ms, and processing times at S, R, and D are negligible.

(a) Suppose $R_1 = R_2 = 10$ Mbps.

   i. Find throughput T as a function of the Window Size W.

   ii. Find the smallest W that maximizes T. Call this value W^*.

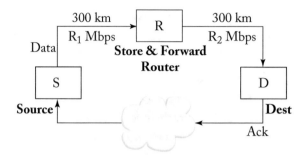

Figure 7.13: Figure for transport Problem 5.

     iii. Find average delay at R for $W = W^*$.

  (b) Suppose $R_1 = 10$ Mbps and $R_2 = 5$ Mbps. Find the average occupancy and average delay at R for $W = W^*$.

P7.6 A good model for analyzing the performance of a TCP congestion control scheme is based on the Figure 7.14 that shows a simplified evolution of window size in the congestion avoidance phase. As shown in the figure, the maximum window size is assumed to be W. It is also known that the Round Trip Time is T and the TCP segment size is M. It can also be assumed that each congestion event resulting in reduction of the window size is due to loss of a single packet. Answer the following questions using this model.

  (a) Calculate the number of round trip times between two loss events.

  (b) Calculate the number of packets transmitted between two loss events.

  (c) Calculate the packet loss rate.

  (d) Calculate the TCP throughput.

  (e) Show that TCP throughput can be expressed as $K/(T \sqrt{q})$, where q is the packet loss rate, and K is some constant.

P7.7 Consider a wireless LAN (WLAN) operating at 11 Mbps that follows the 802.11 MAC protocol with the following parameters: Slot $= 20 \, \mu s$, DIFS $= 2.5$ Slots, SIFS $= 0.5$ Slots, ACK $= 10.5$ Slots (including Physical Layer overhead), Physical Layer overhead for data frames $= 10$ Slots, and Collision Windows grow as 31, 63, 127, ... There is only one user present in this WLAN who wants to transmit a file with 10^6 bytes of data using UDP. Each transmission takes place either with or without channel errors. Recall that transmission of each data or acknowledgment frame incurs the Physical Layer overhead mentioned above. Assume the following:

  • All data frames are 1,100 bytes long which includes 100 bytes of overhead for MAC and upper layers.

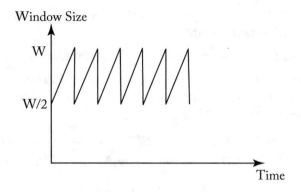

Figure 7.14: Figure for transport Problem 6.

- Propagation delay is negligible.

- An acknowledgment is always received without any errors.

(a) How long will it take to transmit the file on average if data frames are always transmitted without errors?

(b) How long will it take to transmit the file on average if each new data frame is received correctly either after the original transmission or after one retransmission with probabilities 0.9 and 0.1, respectively?

(c) If each data frame transmission (an original transmission or a retransmission) has the probability 0.9 of being received correctly, how will the time to transmit the file compare to the estimate in (b) above? Explain your answer. (Only a qualitative answer is needed for this part.)

P7.8 Suppose there are three aligned nodes; the source A, the relay node B in the middle, and the destination C, equipped with 802.11b wireless LAN radios that are configured to use RTS/CTS for packets of all sizes. The distance between A and B is 750 m, same for B and C; the radios are powerful enough for A, B and B, C to communicate, but communications from A–C must go through B. The transmission from A can still interfere with the reception in C (and conversely). No other nodes operate in the area.

Assumptions and Parameters:

- The packets contain 1,100 bytes of data, the TCP ACKs are 55 bytes long (in both cases, this includes IP and TCP headers).

- Propagation speed is 3×10^8 m/s.

- DIFS $= 50\,\mu$s, SIFS $= 10\,\mu$s.

- The preamble, the physical layer header, the MAC header, and trailer take a combined 200 μs per packet to transmit.

- An ACK, RTS, and CTS each have a 200 μs transmission time.

- The contention window would initially be chosen randomly (with uniform probability) to be from 0–31 slots long; one slot is 20 μs. To simplify, we will assume that A, B, and C always draw 14, 15, and 16, respectively.

- The TCP window is always large enough for the packets to be sent.

- Assume that the MAC backoff counters at A, B, and C are initially at zero.

(a) Suppose A wants to send a single TCP segment to C. What is the total time between the beginning of the transmission and the reception of the TCP ACK at A?

(b) In the scenario above, suppose the MAC layer ACK from B–A gets corrupted (due to excessive noise). Describe how the MAC layer at A reacts to this.

(c) Describe the impact of the loss described in the part (b) on the TCP congestion window size and TCP timeout interval at A.

(d) Suppose A has two TCP segments destined for C. Calculate the total delay from the beginning of the first transmission from A until the reception of the second TCP ACK. Assume that C acknowledges each TCP segment individually.

7.10   REFERENCES

The basic specification of TCP is in [83]. The analysis of AIMD is due to Chiu and Jain (see [25] and [26]). The congestion control of TCP was developed by Van Jacobson in 1988 [49]. See [10] for a discussion of the refinements such as fast retransmit and fast recovery.

CHAPTER 8

Models

The goal of this more mathematically demanding chapter is to explain some recent insight into the operations of network protocols. We start with a review of some basic results on graphs. Queueing is a commonplace occurrence in network operations. In order to better design network protocols, it is important to understand the queueing dynamics. Two other points are noteworthy. First, TCP is an approximation of a distributed algorithm that maximizes the total utility of the active connections. Second, a new class of "backpressure" protocols optimize the scheduling, routing, and congestion control in a unified way.

8.1 GRAPHS

A *directed graph* is a set of *vertices*, or *nodes*, and a set of *arcs*. An arc is a pair of vertices. Figure 8.1 shows a directed graph with vertices $\{1, 2, \ldots, 5\}$ and arcs $\{(1, 2), (2, 3), \ldots, (4, 5), (5, 4)\}$. Each arc (i, j) has a *capacity* $C(i, j)$, which is a positive number that is equal to the maximum rate at which a flow can go through that arc.

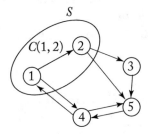

Figure 8.1: A directed graph and a cut S.

A *cut* in a graph is a subset of the nodes. The figure shows cut $S = \{1, 2\}$. One defines the capacity of a cut S as the sum of the capacities of the arcs from S to the complement S^c of S. For instance, in the figure, the *capacity* of the cut S is equal to $C(1, 4) + C(2, 3) + C(2, 5)$. The capacity of a cut is a bound on the rate at which flows can go from S to S^c.

8.1.1 MAX-FLOW, MIN-CUT

Imagine a flow that goes out of some vertex v of a directed graph and uses the arcs to get to some other vertex w in a way that the flow is conserved at every vertex. For instance, in the graph of

Figure 8.1, a flow from 1 to 5 uses the arcs $(1, 2)$ and $(1, 4)$. The flow along $(1, 4)$ continues along $(4, 5)$ and the flow along $(1, 2)$ splits among $(2, 3)$ and $(2, 5)$. The rate of flow $R(1, 2)$ along $(1, 2)$ is the sum $R(2, 3) + R(2, 5)$ of the rates along $(2, 3)$ and along $(2, 5)$. By definition of the capacity of the arcs, it must be that $R(i, j) \leq C(i, j)$ for every arc (i, j).

The *max-flow min-cut* theorem [32] says that the maximum flow from some vertex v to another vertex w is equal to the minimum capacity of the cuts S where $v \in S$ and $w \in S^c$. It is obvious that the rate of flow from v to w cannot exceed the value of any cut S with $v \in S$ and $w \in S^c$. It is a bit more tricky to show that it can be equal to the minimum value of the cuts.

To prove this result, one considers a flow that we think has the maximum possible rate r from v to w but is such that r is at least $\epsilon > 0$ less than the capacity of every cut S with $v \in S$ and $w \in S^c$. We show that this is not possible. The flow from v to w goes along a number of paths. Consider one particular path. If there is some $\delta > 0$ such that each arc along this path has an unused capacity equal to δ, then one can increase the rate of the flow by δ along the path, and we are done. Now, assume the opposite. That is, every path from v to w contains at least one saturated arc. Start with S being the empty set. Follow one of the paths from v to w until you reach a saturated link. For instance, the path consists of links $(1, 2, 3, \ldots, m, \ldots, n)$ and the first saturated arc is m. Add the vertices $\{1, 2, \ldots, m\}$ to the set S. Do this for all the paths from v to w and list the first arcs that are saturated, for instance (i, j, k, l, p, q). The claim is then that the capacity of the cut S is equal to the rate r of the flow. Indeed, the set of arcs (i, j, k, l, p, q) is the set of arcs with source in S and destination in S^c because we explored all the possible paths from v to w. Also, all these arcs are saturated, so that the capacity of each of these arcs is the rate of the flow along the arc. But this is a contradiction since we assumed that the rate of the flow was at least ϵ smaller than the value of every cut.

The proof given above does not provide an algorithm for computing the maximum flow. See [63] for such algorithms.

8.1.2   COLORING AND MAC PROTOCOLS

An *undirected graph* is a set of vertices and a set of undirected *links* that are sets of two different vertices. Figure 8.2 shows two undirected graphs. Two vertices are *neighbors* if they are joined by an arc. The graph is *connected* if all the vertices can be reached from every vertex. A graph is *complete* if any two nodes are connected by a link. The *degree* of a vertex is the number of its neighbors. The *maximum degree* of a graph is the maximum degree of its vertices.

A *coloring* of the graph is an assignment of colors to each vertex so that no two neighboring vertices have the same color. The figure shows a coloring of the graphs. The minimum number of colors required to color a graph is called the *chromatic number* of the graph.

If there is a set of vertices that are pairwise connected, which is called a *clique*, then all these nodes must have a different color. Thus, the chromatic number is at least as large as the largest number of nodes in a clique. The graph in the left-hand part of Figure 8.2 has a clique $\{2, 3, 5\}$ with three nodes. Hence, its chromatic number is at least 3. Note that the largest clique in the

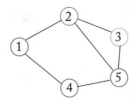 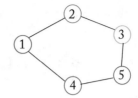

Figure 8.2: Three colored undirected graphs.

graph in the right-hand part of Figure 8.2 has two nodes. However, the chromatic number of the graph is 3. Thus, the maximum number of nodes in a clique is a lower bound on the chromatic number.

Brook's Theorem states that the chromatic number of a connected graph with maximum degree Δ is at most Δ, except if it is an odd cycle or a complete graph, in which case it is equal to $\Delta + 1$. For instance, the graph on the left in Figure 8.2 has maximum degree 3 and it can be colored with three colors. The graph on the right is a cycle, thus its maximum degree is 2, but its chromatic number is 3. Note that the chromatic number of an even cycle is 2. Also, a star graph with Δ branches has maximum degree Δ but its chromatic number is 2. Thus, Brook's upper bound can be quite loose.

Imagine that the vertices of a graph correspond to wireless nodes and that two vertices are connected if the nodes cannot transmit without interfering with one another. In this case, one says that the undirected graph is a *conflict graph* for the wireless nodes. Consider then a coloring of that graph. All the nodes of the same color can transmit without interfering with one another. One says that these nodes form an *independent set*. An independent set is called a *maximal independent set* if every other node is attached to one of the nodes in the set.

One application of coloring is frequency allocation to access points. Say that the vertices in the graphs of Figure 8.2 represent WiFi access points. The colors correspond to distinct channels. The coloring makes sure that the access points do not interfere with one another. A similar application is the allocation of channels to base stations.

Another application is the allocation of time slots in a mesh network. Say that the packets arrive at the wireless nodes that have to transmit them. For simplicity, assume that these packets have to be transmitted once and then leave the network. Let λ_i be the arrival rate of packets at node i, for every $i = 1, 2, \ldots, J$ where J is the total number of nodes. Let also $\lambda = \{\lambda_1, \ldots, \lambda_J\}$. When is a vector λ feasible? That is, when is it possible for the nodes to transmit the arriving packets without interference and keep up with the arrival rates? Since nodes in an independent set can transmit together, let $\{S_1, \ldots, S_K\}$ be all the maximal independent sets and $\mathbf{p} = (p_1, \ldots, p_K)$ be a vector with nonnegative components that add up to one. If the nodes in the independent set S_k transmit a fraction p_k of the time, for $k = 1, \ldots, K$, then node i transmits a fraction of the time equal to $\sum_{k=1}^{K} p_k 1\{i \in S_k\}$. Thus, λ is feasible if and only if there exist some $\mathbf{p}$ with $p_k \geq 0$ for all k and $\sum_{k=1}^{K} p_k = 1$ such that

$$\lambda_i \leq \sum_{k=1}^{K} p_k 1\{i \in S_k\}, i = 1, \ldots, J. \tag{8.1}$$

One centralized implementation of a transmission scheme is then to compute $\mathbf{p}$, then divide time into slots that are just large enough to transmit a packet, choose some number M and allocate periodically, every M time slots, approximately $p_k M$ time slots to the nodes in the independent set S_k, for $k = 1, \ldots, K$. This "time-synchronized mesh protocol" (TSMP) may be usable when the transmission rates λ change very rarely compared to the time needed to recompute $\mathbf{p}$ and reallocate the time slots.

A decentralized scheme to keep up with feasible rates λ is a queue-based carrier sense multiple access protocol (Q-CSMA) that works as follows. When a node has a packet to transmit, it waits until it senses the channel idle. The node then starts a countdown timer with an initial random value that has a mean that is a decreasing function of the channel backlog. The timer counts down when the node senses the channel is idle. When the timer runs out, the packet transmits a packet. All the nodes implement the same scheme. One can show that this scheme keeps up with any λ such that the inequalities (8.1) hold with some positive gap ϵ. Unfortunately, the packet delays increase when ϵ is small. This scheme is combined with an admission control that stops accepting packets in the queue when its backlog exceeds a large value that depends on the priority of the packets. This admission control effectively adjusts the rate λ so that they are feasible and also favor the high priority packets. It can also select rates for which the delays of high-priority packets are acceptable. The advantages of Q-CSMA over TSMP is that it does not require knowing λ and adapts automatically to changing conditions. Moreover, it does not require wasting bandwidth to communicate the time slot allocations. See Section 8.6 for a discussion of this protocol.

8.2   QUEUES

In networking, the goal of queuing theory is to quantify the delays and backlogs of packets. A key observation is that fluctuations in arrivals and in packet lengths are the causes of delays and backlog. For instance, if packets arrive at a transmitter every second and always take less than a second to be transmitted, then no backlog ever builds up and the delays are limited to the transmission times. However, if occasionally packets arrive faster than the transmitter can send them, a backlog and delays build up. Queuing theory provides models of such fluctuations and corresponding results on the backlogs and delays. This section reviews some classical results of queuing theory: the M/M/1 queue and Jackson networks.

In queuing theory, the terminology is that a *customer* arrives at a queue, waits until his time to be served, then faces a service time before he exits the queue. In a queuing network, customers arrive, then go through a sequence of queues. One is interested in the backlog of customers in the queues and the delay of customers from their arrival until their departure.

8.2.1 M/M/1 QUEUE

Section 2.2.6 explained the main results for the M/M/1 queue. This section justifies those results.

Recall the basic assumptions of the M/M/1 queue. The arrivals and service completions are independent, memoryless, and stationary, and customers arrive one at a time. This means that the events that an arrival occurs and that the service in progress completes in the next ts are independent and are also independent of the past; moreover, the probability of these events depends on t but not on the current time. An arrival process with these properties is called a *Poisson process*. Also, services that complete with a probability that is independent of how long the customer has been served so far are *exponentially distributed*. Thus, an M/M/1 queue has Poisson arrivals and exponentially distributed service times. In the notation M/M/1, the first M indicates that the arrivals are memoryless, i.e., Poisson. The second M indicates that the service times are memoryless, i.e., exponentially distributed. Finally, 1 indicates that there is a single server.

Let $0 < \epsilon \ll 1$ and observe the number X_t of packets in an M/M/1 queue every ϵ time units. That is, consider the random sequence of values $X_0, X_\epsilon, X_{2\epsilon}, \ldots$. Because of the basic assumptions, this sequence is a discrete-time Markov chain (see Section 4.5). The transition probabilities of this Markov chain are as follows:

$$P(n, n+1) \approx \lambda\epsilon, \, P(n, n-1) \approx \mu\epsilon, \, P(n, n) \approx 1 - \lambda\epsilon - \mu\epsilon.$$

The approximation signs mean equality up to a term negligible in ϵ. The justification for these values is that an arrival occurs in ϵ seconds with a probability close to $\lambda\epsilon$ and a service completes with a probability close to $\mu\epsilon$. The probability that more than one arrival or service completion occurs in ϵ time units is negligible when $\epsilon \ll 1$.

This Markov chain is irreducible, which means that it can go from any state (a value of X_n) to any other state, possibly in multiple steps. Assume that $\lambda < \mu$, i.e., that the packets arrive more slowly than they can be served, on average. Then, solving the balance equations shows that the invariant probability $\pi(n)$ that there are n packets in the queue is given by

$$\pi(n) = (1 - \rho)\rho^n, n \geq 0$$

where $\rho := \lambda/\mu$. For instance, for $n \geq 1$, we verify that

$$\pi(n)(1 - P(n, n)) = \pi(n-1)P(n-1, n) + \pi(n+1)P(n+1, n),$$

i.e., that

$$\pi(n)(\lambda\epsilon + \mu\epsilon) = \pi(n-1)\lambda\epsilon + \pi(n+1)\mu\epsilon,$$

or, after substituting the expression for $\pi(n)$ and dividing both sides by $(1 - \rho)\epsilon$,

$$\rho^n(\lambda + \mu) = \rho^{n-1}\lambda + \rho^{n+1}\mu.$$

Since $\rho = \lambda/\mu$, we see that this equation is satisfied.

Under this invariant distribution, we find that the average number L of packets in the system is

$$L = \sum_{n \geq 0} n(1 - \rho)\rho^n = \frac{\rho}{1 - \rho} = \frac{\lambda}{\mu - \lambda}. \tag{8.2}$$

Using Little's Result $L = \lambda W$ (see Section 2.2.7), we conclude that the average delay W of a packet in the system is given by

$$W = \frac{1}{\mu - \lambda}. \tag{8.3}$$

8.2.2 JACKSON NETWORKS

A Jackson network is the simplest model of a network of queues. It consists of interconnected M/M/1 queues. Specifically, there are J queues: $1, 2, \ldots, J$. Customers of classes $c = 1, 2, \ldots, C$ arrive at the network independently in a memoryless way, i.e., as Poisson processes with respective rates λ_c. A customer of class c goes through a set $S(c) \subset \{1, \ldots, J\}$ of queues in some specific order. For simplicity, we assume that no customer visits a queue more than once. Thus, the total arrival rate of customers into the network is $\lambda = \sum_{c=1}^{C} \lambda_c$. Also, the total rate of customers that go through some queue j is $\gamma_j := \sum_{c=1}^{C} \lambda_c 1\{j \in S(c)\}$. That is, γ_j is the sum of the arrival rate λ_c of the classes of customers that go through queue j. The service times in queue j are independent of other service times and of the arrivals and are exponentially distributed with mean μ_j, the same for every class of customers. We assume that $\gamma_j < \mu_j$ for all j.

The main results for Jackson networks are as follows. The average delay W_c of a customer of class c, from its arrival into the network until its departure from the network is given by the following expression:

$$W_c = \sum_{j \in S(c)} \frac{1}{\mu_j - \gamma_j}. \tag{8.4}$$

Moreover, the average number L_j of customers in queue j is given by

$$L_j = \frac{\gamma_j}{\mu_j - \gamma_j}. \tag{8.5}$$

Comparing (8.4) with (8.3), we see that a customer faces an average delay $1/(\mu_j - \lambda_j)$ through queue j, as if that queue were an M/M/1 queue. This result is remarkable because in general the queues are not M/M/1 queue as their arrivals may not be memoryless.

Although this result is very easy to state, its derivation is a bit too long for this short book.

8.2.3 QUEUING VS. COMMUNICATION NETWORKS

In a communication network, the service time of a packet is equal to its length divided by the rate of the transmitter. Thus, if the lengths of the packets are independent and exponentially distributed with mean B bits and the transmitter rate is R bits per second, then the service times are independent and exponentially distributed with mean $\mu^{-1} = B/R$ s. If, in addition, the packets arrive as a Poisson process with rate λ, then the system is an M/M/1 queue and we know its average backlog and the average delay of packets through the queue.

Assume that upon leaving the queue the packets arrive at a second queue that stores them and starts transmitting them when they are complete. The key observation is that, at the second queue, the arrival times are not independent of the service times. For instance, the time between the arrival times of packet n and packet $n + 1$ cannot be smaller than the length of packet n divided by the service rate R of the first queue. Consequently, the arrival time of packet $n + 1$ depends on the service time of packet n in the second queue. Thus, the system of two queues in tandem is not a Jackson network and the average delay of packets in this system is not given by the formula for a delay in a Jackson network. Figure 8.3 compares the average delay in the communication network, as found by simulations, and that in a Jackson network. The comparison shows that the differences are rather small. Hence, it may be justified to use the Jackson network model to predict delays in a communication network and we use that approximation in the remainder of this section.

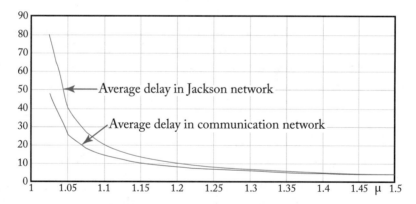

Figure 8.3: Approximating a communication network with a Jackson network.

As an application of the results for Jackson networks, consider the problem of balancing the traffic in a network. Say that there are two possible paths for packets from their source to their destination. It is plausible that one might reduce the delay of these packets by splitting the traffic between the two paths. As a simple model of this situation, say that the packets arrive as a Poisson process with rate λ and can go through two queues with respective service rates μ_1 and μ_2. We choose to send a fraction p of the packets through the first queue and the others through the second queue. If each packet is selected independently with probability p to go to the first queue and to the second queue otherwise, then the arrivals at the first queue are memoryless and have rate λp whereas those at the second queue are memoryless and have rate $\lambda(1 - p)$. Thus, the two queues are M/M/1 queues and their average delays are $1/(\mu_1 - \lambda p)$

and $1/(\mu_2 - \lambda(1 - p))$, respectively, assuming that $\lambda p < \mu_1$ and $\lambda(1 - p) < \mu_2$. Since a packet experiences a delay in the first queue with probability p and in the second queue with probability $1 - p$, the average delay T per packet is given by

$$T = \frac{p}{\mu_1 - \lambda p} + \frac{1 - p}{\mu_2 - \lambda(1 - p)}.$$

Assuming that $\lambda < \mu_1 + \mu_2$, we can find the value of p that minimizes the average delay T by setting to zero the derivative with respect to p of the expression above. Algebra shows that the minimizing value is

$$p = \frac{\lambda\sqrt{\mu_1} + \mu_1\sqrt{\mu_2} - \mu_2\sqrt{\mu_1}}{\lambda(\sqrt{\mu_1} + \sqrt{\mu_2})}$$

For instance, if $\mu_1 = \mu_2$, then $p = 1/2$, which is intuitively clear. As another example, if $\lambda = 1 = \mu_1$ and $\mu_2 = 2$, then $p \approx 0.17$. In this case, we find that the average delay in the first queue is $1/(\mu_1 - \lambda p) \approx 1.2$ whereas the average delay in the second queue is $1/(\mu_2 - \lambda(1 - p)) \approx 0.85$. Thus, somewhat surprisingly, the optimal load balancing does not equalize the average delays in the two queues. In this example, the packets that happen to go through the first queue face a larger average delay than the other packets.

One might worry that sending packets along two paths that do not have the same average delay might cause havoc for TCP, and one would be correct. To avoid this problem, packets from a source to a destination are split in a way that the packets of the same connection, i.e., the same four-tuple (IP source, TCP source, IP destination, TCP destination) follow the same paths.

8.3   THE ROLE OF LAYERS

Protocols regulate the delivery of information in a network by controlling errors, congestion, routing, and the sharing of transmission channels. As we learned in Chapter 2, the original view was that these functions are performed by different layers of the network: layer 4 for error and congestion control, 3 for routing, and 2 for the medium access control. This layer decomposition was justified by the structure that it imposes on the design and, presumably, because it simplifies the solution of the problem.

However, some suspicion exists that the "forced" decomposition might result in a loss of efficiency. More worrisome is the possibility that the protocols in the different layers might interact in a less than constructive way. To address this interaction of layers, some researchers even proposed some "cross-layer" designs that sometimes resembled strange Rube Goldberg contraptions.

Recently, a new understanding of control mechanisms in networks emerged through a series of remarkable papers. Before that work, schemes like TCP or CSMA/CA seemed to be clever but certainly ad hoc rules for controlling packet transmissions: slow down if the network seems congested. For many years, nobody really thought that such distributed control rules were approaching optimal designs. The control community by and large did not spend much time exploring TCP.

The new understanding starts by formulating a global objective: maximize the total utility of the flows in the network. The next step is to analyze the problem and show that, remarkably, it decomposes into simpler problems that can be thought of as different protocols such as congestion control and routing. One might argue that there is little point in such after-the-fact analysis, even though the results of the analysis yield some surprises: improved protocols that increase the utility of the network. Of course, it may be a bit late to propose a new TCP or a new BGP. However, the new protocols might finally produce multi-hop wireless networks that work. Over the last decade, researchers have been developing multiple heuristics for routing, scheduling, and congestion control in multi-hop wireless networks. They have been frustrated by the poor performance of those protocols. A common statement in the research community is that "three hops = zero throughput." The protocols derived from the theoretical analysis hold the promise of breaking this logjam. Moreover, the insight that the results provides is valuable because it shows that protocols do not have to look like fairly arbitrary rules but can be derived systematically.

8.4   CONGESTION CONTROL

Chapter 7 explained the basic mechanism that the Internet uses to control congestion. Essentially, the source slows down when the network gets congested. We also saw that additive increase, multiplicative decrease has a chance to converge to a fair sharing of one link by multiple connections. A major breakthrough in the last ten years has been a theoretical understanding of how it is possible for a distributed congestion control protocol to achieve such a fair sharing in a general network. We explain that understanding in this section.

8.4.1   FAIRNESS VS. THROUGHPUT

Figure 8.4 shows a network with three flows and two nodes A and B. Link a from node A to node B has capacity 1 and so does link b out of node B. We denote the rates of the three flows by $x_1, x_2,$ and x_3. The feasible values of the rates (x_1, x_2, x_3) are nonnegative and such that

$$g_a(\mathbf{x}) := x_1 + x_2 - 1 \leq 0 \text{ and } g_b(\mathbf{x}) := x_1 + x_3 - 1 \leq 0.$$

For instance, $(0, 1, 1)$ is feasible and one can check that these rates achieve the maximum value 2 of the total throughput $x_1 + x_2 + x_3$ of the network. Of course, this set of rates is not fair to the user of flow 1 who does not get any throughput. As another example, the rates $(1/2, 1/2, 1/2)$ are as fair as possible but achieve only a total throughput equal to 1.5. This example illustrates that quite typically there is a tradeoff between maximizing the total throughput and fairness.

To balance these objectives, let us assign a *utility* $u(x_i)$ to each flow i ($i = 1, 2, 3$), where $u(x)$ is a concave increasing function of $x > 0$. The utility $u(x)$ reflects the value to the user of

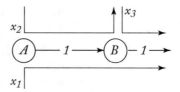

Figure 8.4: Three flows in a simple network.

the rate x of the connection. The value increases with the rate. The concavity assumption means that there is a diminishing return on larger rates. That is, increasing the rate by 1 Kbps is less valuable when the rate is already large.

We choose the rates $\mathbf{x} = (x_1, x_2, x_3)$ that maximize the sum $f(\mathbf{x}) = u(x_1) + u(x_2) + u(x_3)$ of the utilities of the flows. For instance, with $\alpha \geq 0$, say that

$$u(x) = \begin{cases} \frac{1}{1-\alpha} x^{1-\alpha}, \text{ for } \alpha \neq 1 \\ \log(x), \text{ for } \alpha = 1. \end{cases} \tag{8.6}$$

For $\alpha \neq 1$, the derivative of this function is $x^{-\alpha}$ which is positive and decreasing, so that the function is increasing and concave. For $\alpha = 1$, the function $\log(x)$ is also increasing and concave.

That is, for $\alpha \neq 1$, we look for the rates $\mathbf{x}$ that solve the following problem:

$$\text{Maximize } f(\mathbf{x}) := \frac{1}{1-\alpha} \left[x_1^{1-\alpha} + x_2^{1-\alpha} + x_3^{1-\alpha} \right]$$
$$\text{subject to } x_1 + x_2 \leq 1 \text{ and } x_1 + x_3 \leq 1. \tag{8.7}$$

Since $f(\mathbf{x})$ is increasing in each x_i, the solution must be such that $x_1 + x_2 = 1 = x_1 + x_3$. Let $x = x_1$ so that $x_2 = x_3 = 1 - x$. With this notation,

$$f(\mathbf{x}) = \frac{1}{1-\alpha} \left[x^{1-\alpha} + 2(1-x)^{1-\alpha} \right].$$

The value of x that maximizes that expression is such that the derivative with respect to x is equal to zero. That is,

$$x^{-\alpha} - 2(1-x)^{-\alpha} = 0.$$

Hence,

$$x = 2^{-1/\alpha}(1-x), \text{ so that } x = \frac{1}{1 + 2^{1/\alpha}}.$$

Note that $(x, 1-x, 1-x)$ goes from the maximum throughput rates $(0, 1, 1)$ to the perfectly fair rates $(1/2, 1/2, 1/2)$ as α goes from 0 to ∞.

Figure 8.5 shows the values of x_1, x_2, x_3 that maximize the sum of utilities as a function of α. As the graphs show, as α goes from 0 to ∞, the corresponding rates go from the values that maximize the sum of the rates to those that maximize the minimum rate.

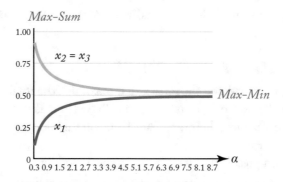

Figure 8.5: Rates that maximize the sum of utilities as a function of α.

With the specific form of the function $u(\cdot)$ in (8.6), the maximizing rates $\mathbf{x}$ are said to be α-fair. For $\alpha = 0$, the objective is to maximize the total throughput, since in that case $u(x) = x$ and $f(\mathbf{x}) = x_1 + x_2 + x_3$.

For $\alpha \to \infty$, the rates that maximize $f(\mathbf{x})$ maximize the minimum of the rates $\{x_1, x_2, x_3\}$. To see this, note that $u'(x) = x^{-\alpha} \gg u'(y) = y^{-\alpha}$ if $x < y$ and $\alpha \gg 1$. Consequently, if $x_1 < x_2, x_3$, then one can increase $f(\mathbf{x})$ by replacing x_2 by $x_2 - \epsilon$, x_3 by $x_3 - \epsilon$ and x_1 by $x_1 + \epsilon$ for $\epsilon \ll 1$. This modification does not violate the capacity constraints but increases $f(\mathbf{x})$ by $\epsilon[u'(x_1) - u'(x_2) - u'(x_3)]$, which is positive if α is large enough.

For a general network, this argument shows that if $\alpha \gg 1$, the rates $\mathbf{x}$ that maximize $f(\mathbf{x})$ must be such that it is not possible to increase some x_i without decreasing some x_j that is smaller than x_i. Rates with that property are called the *max-min fair* rates. To see why only max-min rates can maximize $f(\mathbf{x})$ when $\alpha \gg 1$, assume that one can replace x_i by $x_i + \epsilon$ by only decreasing some x_j's that are larger than x_i. In that case, the previous argument shows that the net change of $f(\mathbf{x})$ is positive after such a change, which would contradict the assumption that the rates maximize the function.

For the case of $\alpha = 1$, we make the following observation for a general network. Suppose $\mathbf{x}^* = (x_1^*, \ldots, x_n^*)$ maximizes $u(x_1) + \cdots + u(x_n)$ subject to constraints. Say that $\mathbf{x} = (x_1, \ldots, x_n)$ is a vector close to $\mathbf{x}^*$ that also satisfies the constraints. Then, $u(x_1) + \cdots + u(x_n) \leq u(x_1^*) + \cdots + u(x_n^*)$. Also, since $\mathbf{x}$ and $\mathbf{x}^*$ are close, $u(x_m) \approx u(x_m^*) + u'(x_m^*)dx_m$ where $dx_m = x_m - x_m^*$. Consequently, $u'(x_1^*)dx_1 + \cdots + u'(x_n^*)dx_n \leq 0$. If $u(x) = \log(x)$, one has $u'(x) = 1/x$. In that case, $dx_1/x_1^* + \cdots + dx_n/x_n^* \leq 0$. Thus, an increase of x_1 by 1% corresponds to changes in the other values x_m so that the sum of their relative changes is less than -1%. One says that $\mathbf{x}^*$ is *proportionally fair*.

The example considered in this section shows that one can model the fair allocation of rates in a network as those rates that maximize the total utility of the users subject to the capacity constraints imposed by the links. Moreover, by choosing α, one adjusts the tradeoff between efficiency (maximizing the sum of throughputs) and strict fairness (maximizing the utility of the worst-off user).

8.4.2   DISTRIBUTED CONGESTION CONTROL

In the previous section, we calculated the rates $\mathbf{x}$ that maximize the total utility $f(\mathbf{x})$ assuming a complete knowledge of the network. In practice, this knowledge is not available anywhere. Instead, the different sources control their rates based only on their local information. We consider that distributed problem next.

We use the following result that we explain in the appendix.

Theorem 8.1   *Let $f(\mathbf{x})$ be a concave function and $g_j(\mathbf{x})$ be a convex function of $\mathbf{x}$ for $j = a, b$. Then $\mathbf{x}^*$ solves the following problem, called the* primal problem,

$$Maximize\ f(\mathbf{x})$$
$$subject\ to\ g_j(\mathbf{x}) \leq 0,\ j = a, b \tag{8.8}$$

if and only if x^* satisfies the constraint (8.8) and

$$\mathbf{x}^*\ maximizes\ L(\mathbf{x}, \lambda^*) := f(\mathbf{x}) - \lambda_a^* g_a(\mathbf{x}) - \lambda_b^* g_b(\mathbf{x}), \tag{8.9}$$

for some $\lambda_a^* \geq 0$ and $\lambda_b^* \geq 0$ such that

$$\lambda_j^* g_j(\mathbf{x}^*) = 0,\ for\ j = a, b. \tag{8.10}$$

Moreover, if

$$L(\mathbf{x}(\lambda), \lambda) = \max_{\mathbf{x}} L(\mathbf{x}, \lambda),$$

then the variables λ_a^* and λ_b^* minimize

$$L(\mathbf{x}(\lambda), \lambda). \tag{8.11}$$

The function L is called the *Lagrangian* and the variables $(\lambda_a^*, \lambda_b^*)$ the *Lagrange multipliers* or *shadow prices*. The relations (8.10) are called the *complementary slackness conditions*. The problem (8.11) is called the *dual* of (8.8).

To apply this result to problem (8.7) with a general utility function $u(x)$ and the two links a and b, we first compute the Lagrangian:

$$\begin{aligned} L(\mathbf{x}, \lambda) &= f(\mathbf{x}) - \lambda_a g_a(\mathbf{x}) - \lambda_b g_b(\mathbf{x}) \\ &= \{u(x_1) - (\lambda_a + \lambda_b)x_1\} + \{u(x_2) - \lambda_a x_2\} + \{u(x_3) - \lambda_b x_3\} + (\lambda_a + \lambda_b). \end{aligned}$$

Second, we find the value $\mathbf{x}(\lambda)$ of $\mathbf{x}$ that maximizes $L(\mathbf{x}, \lambda)$ for a fixed value of $\lambda = (\lambda_a, \lambda_b)$. Since $L(\mathbf{x}, \lambda)$ is the sum of three functions, each involving only one of the variables x_i, to maximize this sum, the value of x_1 maximizes

$$u(x_1) - (\lambda_a + \lambda_b)x_1.$$

Similarly, x_2 maximizes

$$u(x_2) - \lambda_a x_2$$

and x_3 maximizes

$$u(x_3) - \lambda_b x_3.$$

The interpretation is that each link $j = a, b$ charges the user of each flow a price λ_j per unit rate. Thus, the user of flow 1 pays $x_1(\lambda_a + \lambda_b)$ because that flow goes through the two links. That user chooses x_1 to maximize the *net utility* $u(x_1) - (\lambda_a + \lambda_b)x_1$. Similar considerations apply to flows 2 and 3.

The key observation is that the maximization of L is decomposed into a separate maximization for each user. The coupling of the variables **x** in the original problem (8.7) is due to the constraints. The maximization of L is unconstrained and decomposes into separate problems for each of the variables. That decomposition happens because the constraints are linear in the flows, so that each variable x_i appears in a different term in the sum L. Note also that the price that each link charges is the same for all its flows because the constraints involve the sum of the rates of the flows. That is, the flows are indistinguishable in the constraints.

To find the prices λ^*, one uses a *gradient algorithm* to minimize $L(\mathbf{x}, \lambda)$:

$$\lambda_j(n+1) = \lambda_j(n) - \beta \frac{d}{d\lambda_j} L(\mathbf{x}(\lambda), \lambda) = \lambda_j(n) + \beta g_j(\mathbf{x})$$

where **x** is the vector of rates at step n of the algorithm and β is a parameter that controls the *step size* of the algorithm. This expression says that, at each step n, the algorithm adjusts λ in the direction opposite to the gradient of the function $L(\mathbf{x}(\lambda), \lambda)$, which is the direction of steepest descent of the function. For the new value of the prices, the users adjust their rates so that they approach $\mathbf{x}(\lambda)$ for the new prices. That is, at step n user 1 calculates the rate $x_1(n)$ as follows:

$$x_1(n) \ \text{maximizes} \ u(x_1) - (\lambda_a(n) + \lambda_b(n))x_1,$$

and similarly for $x_2(n)$ and $x_3(n)$.

Figure 8.6 illustrates the gradient algorithm. The figure shows that the algorithm adjusts **x** in the direction of the gradient of $L(\mathbf{x}, \lambda)$ with respect to **x**, then λ in the direction opposite to the gradient of $L(\mathbf{x}, \lambda)$ with respect to λ. This algorithm searches for the *saddle point* indicated by a star in the figure. The saddle point is the minimum over λ of the maximum over **x** of $L(\mathbf{x}, \lambda)$.

To see how the nodes can calculate the prices, consider the queue length $q_j(t)$ at node j at time t. This queue length increases at a rate equal to the difference between the total arrival rate of the flows minus the service rate. Thus, over a small interval τ,

$$q_j((n+1)\tau) = q_j(n\tau) + \tau g_j(\mathbf{x})$$

where **x** is the vector of rates at time $n\tau$. Comparing the expressions for $\lambda_j(n+1)$ and $q_j((n+1)\tau)$, we see that if the steps of the gradient algorithm for λ_j are executed every τ seconds, then

$$\lambda_j \approx (\beta/\tau)q_j.$$

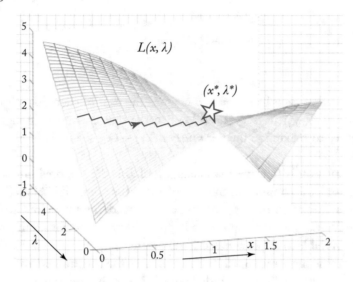

Figure 8.6: The gradient algorithm for the dual problem.

That is, the price of the link should be chosen proportional to the queue length at that link. The intuition should be clear: if queue j builds up, then the link should become more expensive to force the users to reduce their rate. Conversely, if queue j decreases, then the price of link j should decrease to encourage the users to increase their usage of that link.

8.4.3   TCP REVISITED

The TCP algorithm does something qualitatively similar to the algorithm discussed above. To see this, think of the loss rate experienced by a flow as the total price for the flow. This price is the sum of the prices (loss rates) of the different links. The loss rate increases with the queue length, roughly linearly if one uses Random Early Detection (RED) which probabilistically drops packets when the queue length becomes large. A source slows down when the price increases, which corresponds to readjusting the rate x to maximize $u(x) - px$ when the price p increases.

We compare the behavior of TCP's congestion algorithm with the algorithm based on the optimization formulation on the network shown in Figure 8.4.

As we explained in Section 7.5, TCP uses the additive increase, multiplicative decrease (AIMD) scheme.

Figure 8.7 sketches how the average rates of the three connections change over time when the sources use AIMD.

The figure assumes the link capacities $C_a = 8$, $C_b = 4$, and that the round-trip time is twice as long for connection 1 as it is for connections 2 and 3. The figure shows that the long-

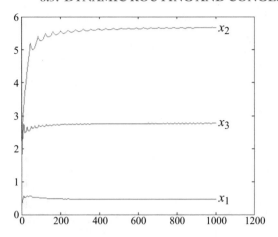

Figure 8.7: Evolution of rates using the algorithm for proportional fairness (where $u(x) = \log(x)$). The figure shows the rates x_1, x_2, x_3 of the three connections, on the Y-axis, as a function of time, on the X-axis, as the algorithm adjusts these rates. The limiting rates show that AIMD is severely biased against connections with a larger round-trip time.

term average rates of the three connections are

$$x_1 = 0.47, \ x_2 = 5.65, \ x_3 = 2.77.$$

Thus, the sum of the average rates for the two links are 6.12 and 3.24, respectively, which is about 80% of the capacities of the two links. The loss in throughput in our model comes from the fact that the buffers are too small and it disappears if one uses larger buffers, which is done in practice.

We see that connection 0 has a very small throughput, partly because it uses two links and partly because its long round-trip time results in slower increases in the rate of its connection, which allows the other connections to take better advantage of available capacity.

Thus, AIMD is a clever ad hoc scheme that stabilizes the rates of the connections so that no router gets saturated, and this was the main goal of its inventor Van Jacobson. However, the scheme results in connection rates that depend strongly on the round-trip times and on the number of links they go through.

When using the dual algorithm, the rates of the three connections adjust, as shown in Figure 8.8. This algorithm converges to the proportionally fair allocation $x_1 = 1.7, x_2 = 6.3, x_3 = 2.3$.

8.5   DYNAMIC ROUTING AND CONGESTION CONTROL

This section illustrates how the utility maximization framework can naturally lead to distributed protocols in different layers.

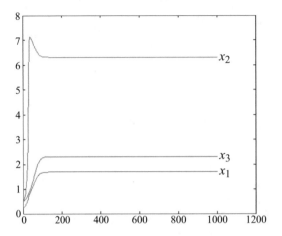

Figure 8.8: Evolution of rates using the proportional fair algorithm. The figure shows the rates x_1, x_2, x_3 of the three connections, on the Y-axis, as a function of time, on the X-axis, as the algorithm adjusts these rates. The limiting rates are proportionally fair.

The network in Figure 8.9 shows four nodes and their backlog $X_1, \ldots, X_4$. Two flows arrive at the network with rates R_a and R_b that their sources control. Node 1 sends bits to node 2 with rate S_{12} and to node 3 with rate S_{13}. The sum of these rates must be less than the transmission rate C_1 of node 1. Similarly, the rate S_{24} from node 2 to node 4 cannot exceed the rate C_2 of the transmitter at node 2 and S_{34} cannot exceed C_3.

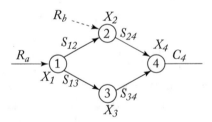

Figure 8.9: A network with two flows.

The objective is to maximize the sum of the utilities

$$u_a(R_a) + u_b(R_b)$$

of the flows while preventing the backlogs from increasing excessively where the functions u_a and u_b are increasing and concave, reflecting the fact that users derive more utility from higher transmission rates, but with diminishing returns in the sense that an additional unit of rate is less valuable when the rate is large than when it is small. For instance, one might have

$$u_j(x) = k_j \frac{1}{1-\alpha} x^{1-\alpha}, j = a, b,$$

for some $\alpha > 0$ with $\alpha \neq 1$ and $k_j > 0$.

To balance the maximization of the utilities while keeping the backlogs small, we choose the rate $(R_a, R_b, S_{12}, S_{13}, S_{24}, S_{34})$ to maximize, for some $\beta > 0$,

$$\phi := \beta[u_a(R_a) + u_b(R_b)] - \frac{d}{dt}\left[\frac{1}{2}\sum_{i=1}^{4} X_i^2(t)\right].$$

To maximize this expression, one chooses rates that provide a large utility $u_a(R_a) + u_b(R_b)$ and also a large decrease of the sum of the squares of the backlogs. The parameter β determines the tradeoff between large utility and large backlogs. If β is large, then one weighs the utility more than the backlogs.

Now,

$$\frac{d}{dt}\left[\frac{1}{2}X_2^2(t)\right] = X_2(t)\frac{d}{dt}X_2(t) = X_2(t)[R_b + S_{12} - S_{24}]$$

because the rate of change of $X_2(t)$ is the arrival rate $R_b + S_{12}$ into node 2 minus the departure rate S_{24} from that node. The terms for the other backlogs are similar. Putting all this together, we find

$$
\begin{aligned}
\phi &= \beta[u_a(R_a) + u_b(R_b)] - X_1[R_a - S_{12} - S_{13}] - X_2[R_b + S_{12} - S_{24}] \\
&\quad - X_3[S_{13} - S_{34}] - X_4[S_{24} + S_{34} - C_4].
\end{aligned}
$$

We rearrange the terms of this expression as follows:

$$
\begin{aligned}
\phi &= [\beta u_a(R_a) - X_1 R_a] + [\beta u_b(R_b) - X_2 R_b] \\
&\quad + S_{12}[X_1 - X_2] + S_{13}[X_1 - X_3] + S_{24}[X_2 - X_4] + S_{34}[X_3 - X_4] + C_4 X_4.
\end{aligned}
$$

The maximization is easy if one observes that the different terms involve distinct variables. Observe that the last term does not involve any decision variable. One finds the following:

$$R_a \text{ maximizes } \beta u_a(R_a) - X_1 R_a$$
$$R_b \text{ maximizes } \beta u_b(R_b) - X_2 R_b$$
$$S_{12} = C_1 \times 1\{X_1 > X_2 \text{ and } X_2 < X_3\}$$
$$S_{13} = C_1 \times 1\{X_1 > X_3 \text{ and } X_3 < X_2\}$$
$$S_{24} = C_2 \times 1\{X_2 > X_4\}$$
$$S_{34} = C_3 \times 1\{X_3 > X_4\}.$$

Thus, the source of flow a adjusts its rate R_a based on the backlog X_1 in node 1, and similarly for flow b. The concavity of the functions u_j implies that R_a decreases with X_1 and R_b with X_2. For the specific forms of the functions indicated above, one finds $R_a = [k_a/X_1]^{1/\alpha}$ and $R_b = [k_b/X_2]^{1/\alpha}$. Node 1 sends the packets to the next node (2 or 3) with the smallest backlog, provided that that backlog is less than X_1; otherwise, node 1 stops sending packets. Node 2 sends packets as long as its backlog exceeds that of the next node and similarly for node 3.

Since the rates R_a and R_b go to zero as X_1 and X_2 increase, one can expect this scheme to make the network stable, and this can be proved. In a real network, there are delays in the feedback about downstream backlogs, and the nodes send packets instead of bits. Nevertheless, it is not hard to show that these effects do not affect the stability of the network.

This mechanism is called a *backpressure algorithm*. Note that the congestion control is based only on the next node. This contrasts with TCP that uses loss signals from all the nodes along the path. Also, the routing is based on the backlogs in the nodes. Interestingly, the control at each node is based only on local information about the possible next nodes downstream. This information can be piggy-backed on the reverse traffic. Thus, this algorithm can be easily implemented in each node without global knowledge of the whole network.

8.6   WIRELESS

What about wireless nodes? In a wired network, one link sees the backlog in the buffer attached to it. However, in a wireless network, if two nodes share some channel, they are not aware of the total backlog of that channel. How can we design a distributed algorithm? We explain an algorithm on a simple example shown in Figure 8.10. Two nodes share one wireless channel. We assume that the utility of flow 1 is $\log(x)$ and that of flow 2 is $2\log(y)$.

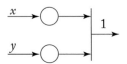

Figure 8.10: Two flows share a wireless channel with capacity 1.

The optimization problem is then:

$$\text{Maximize } f(x, y) = \log(x) + 2\log(y)$$
$$\text{subject to } x + y \leq 1.$$

If we use the dual algorithm approach that we explained above, we find that the nodes should adjust x and y based on the *total* backlog in front of that channel. However, no node sees that total backlog since it is split in two different buffers: one in each node.

At this point, one could ask the two nodes to exchange their backlog information. However, this approach is complex, especially in a large network.

Another approach is feasible, even though it looks rather indirect at first. Let p_1 and p_2 be the fractions of time that the two nodes use the channel. We then formulate the problem as follows:

$$\text{Maximize } f(x, y) = \log(x) + 2\log(y)$$
$$\text{subject to } x \le p_1$$
$$\text{and } y \le p_2$$
$$\text{and } p_1 + p_2 \le 1.$$

To derive the desired distributed algorithm, we further modify the problem as indicated below:

$$\text{Maximize } f(x, y) = \log(x) + 2\log(y) + \beta\phi(p_1) + \beta\phi(p_2)$$
$$\text{subject to } x \le p_1$$
$$\text{and } y \le p_2$$
$$\text{and } p_1 + p_2 = 1.$$

In the objective function, $\phi(z)$ is some bounded concave increasing function of z. If β is small, this is a very small modification of the objective function, so that we have not really changed the problem. We see later what these terms do.

Next, we replace the two inequality constraints by penalties. We keep the equality $p_1 + p_2 = 1$ for later.

$$\begin{aligned} \text{Maximize } h(x, y, p; \lambda) \quad = \quad & \log(x) + 2\log(y) \\ &+ \beta\phi(p_1) + \beta\phi(p_2) - \lambda_1(x - p_1) \\ &- \lambda_2(y - p_2). \end{aligned}$$

We then maximize h over (x, y, p) and minimize over λ using a gradient step. This gives

$$\frac{1}{x} = \lambda_1$$
$$\frac{2}{y} = \lambda_2$$
$$\beta\phi'(p_1) + \lambda_1 = 0$$
$$\beta\phi'(p_2) + \lambda_2 = 0$$
$$\lambda_1(n + 1) = \lambda_1(n) + \gamma(x(n) - p_1(n))$$
$$\lambda_2(n + 1) = \lambda_2(n) + \gamma(y(n) - p_2(n)).$$

As before, the last two identities show that λ_k is proportional to the backlog in node k, for $k = 1, 2$. Thus, the first two equations explain the flow control in terms of the backlogs. The third and fourth equations are novel. They show that

$$p_1 = \psi(-\beta^{-1}\lambda_1) \text{ and } p_1 = \psi(-\beta^{-1}\lambda_2).$$

In this expression, ψ is the inverse of ϕ'. Choosing ϕ is equivalent to choosing ψ. Many such functions result in approximately optimal solutions in terms of throughput but may result in different delays. The key point is that p_1 should increase with the backlog of node 1 and p_2 with the backlog of node 2. Recall that we want $p_1 + p_2 = 1$.

One distributed way to implement these expressions for p_1 and p_2 is CSMA: let node i try to access an idle channel with probability a_i during a "mini" contention slot. Then, if the a_i are small so that we can neglect $a_1 a_2$, we see that node i grabs the channel with probability $a_i/(a_1 + a_2)$. Thus, a_i should be increasing in the backlog of node i. CSMA takes care of $p_1 + p_2 = 1$.

We end up with the following algorithm:

• each node attempts to transmit with a probability that increases with its backlog;

• each node admits new packet with a rate that decreases with its backlog.

This corresponds to a distributed protocol.

Figure 8.11 shows how the algorithm performed. In this example, we used

$$a_1 = \exp\{2Q(1,n)\} \text{ and } a_2 = \exp\{2Q(2,n)\}.$$

There is an art to choosing these functions. Different choices result in different backlogs and, consequently, delays.

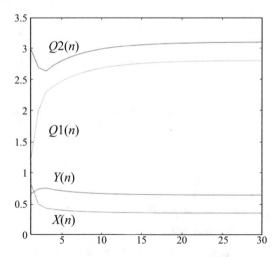

Figure 8.11: Performance of the adaptive CSMA algorithm.

The same approach extends to a general multi-hop wireless network and can be combined with the backpressure routing and flow control.

8.7   APPENDIX: JUSTIFICATION FOR PRIMAL-DUAL THEOREM

In this section, we justify Theorem 8.1.

Consider the Problem (8.8) and assume that the solution is $\mathbf{x}^*$. Assume that $g_a(\mathbf{x}^*) = 0$ and $g_b(\mathbf{x}^*) = 0$, so that both constraints are just satisfied. Since $\mathbf{x}^*$ maximizes f subject to the constraints $g_a \leq 0$ and $g_b \leq 0$, it is not possible to move slightly away from $\mathbf{x}^*$ so as to increase $f(\mathbf{x})$ without increasing the function $g_a(\mathbf{x})$ or $g_b(\mathbf{x})$. That is, it is not possible to move from $\mathbf{x}^*$ in the direction of the gradient $\nabla f(\mathbf{x}^*)$ without also moving in the direction of a gradient $\nabla g_a(\mathbf{x}^*)$ or $\nabla g_b(\mathbf{x}^*)$.

This condition is satisfied in the left part of Figure 8.12, but not in the right part, where $\mathbf{x}$ increases f beyond $f(\mathbf{x}^*)$ without increasing g_a nor g_b. The figure shows that the condition implies that $\nabla f(\mathbf{x}^*)$ must be a nonnegative linear combination of $\nabla g_a(\mathbf{x}^*)$ and $\nabla g_b(\mathbf{x}^*)$.

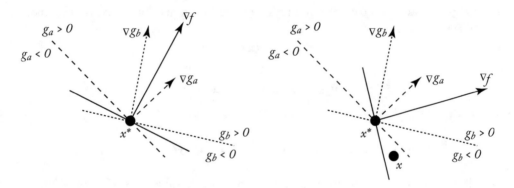

Figure 8.12: Solution and gradients.

Thus, a necessary condition for optimality of $\mathbf{x}^*$ is that

$$\nabla f(\mathbf{x}^*) = \lambda_a \nabla g_a(\mathbf{x}^*) + \lambda_b \nabla g_b(\mathbf{x}^*)$$

for some $\lambda_a \geq 0$ and $\lambda_b \geq 0$ and, moreover,

$$\lambda_a g_a(\mathbf{x}^*) = 0 \text{ and } \lambda_b g_b(\mathbf{x}^*) = 0.$$

(Note that this argument is not valid if the gradients $\nabla g_a(\mathbf{x}^*)$ and $\nabla g_b(\mathbf{x}^*)$ are proportional to each other. Thus, technically, the theorem requires that this does not happen. These conditions are called the *Slater conditions*.)

Finally, we want to show that λ^* minimizes $D(\lambda) := \max_{\mathbf{x}} L(\mathbf{x}, \lambda)$. By definition, $D(\lambda)$ is the maximum of linear functions of λ (one for each value of $\mathbf{x}$). Figure 8.13 shows that this maximum is convex. For simplicity of the figure, we have assumed that there is only one con-

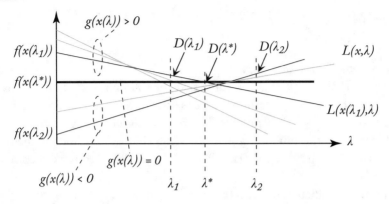

Figure 8.13: The optimal Lagrange multiplier minimizes $D(\lambda)$.

straint $g(\mathbf{x}) \leq 0$, and λ is the corresponding Lagrange multiplier. Note that if $\mathbf{x}(\lambda)$ maximizes $L(\mathbf{x}, \lambda)$ where,

$$L(\mathbf{x}, \lambda) = f(\mathbf{x}) - \lambda g(\mathbf{x}),$$

then

$$\nabla_{\mathbf{x}} L(\mathbf{x}, \lambda) = 0 \text{ for } \mathbf{x} = \mathbf{x}(\lambda).$$

Consequently,

$$\frac{d}{d\lambda} L(\mathbf{x}(\lambda), \lambda) = \nabla_{\mathbf{x}} L(\mathbf{x}(\lambda), \lambda) \cdot \frac{d\mathbf{x}(\lambda)}{d\lambda} + \frac{\partial}{\partial \lambda} L(\mathbf{x}(\lambda), \lambda) = -g(\mathbf{x}(\lambda)).$$

Using this observation, the figure shows that the largest value $f(\mathbf{x}(\lambda^*))$ of $f(\mathbf{x}(\lambda))$ with $g(\mathbf{x}(\lambda)) \leq 0$ corresponds to $g(\mathbf{x}(\lambda^*)) = 0$ and to the minimum value $D(\lambda^*)$ of $D(\lambda)$. The figure is drawn for the case $g(\mathbf{x}(0)) > 0$. The case $g(\mathbf{x}(0)) < 0$ is similar, except that all the lines have a positive slope and $\lambda^* = 0$ then maximizes $D(\lambda)$. For details, the reader is referred to [20].

8.8   SUMMARY

The chapter explains that layers can be understood as solving subproblems of a global optimization problem.

- In the Internet and associated networks, different layers address separately congestion control, routing, and scheduling. Each layer implicitly assumes that the layers below it have solved their problems and have come up with a stable solution. Thus, routing assumes that the links are stable. Congestion control assumes that routing is stable.

- Each layer implements an approximate solution. TCP uses a reasonable AIMD heuristic. Routing looks for a shortest path where the link metrics do not depend on the traffic load. Scheduling may use WFQ with fixed weights.

- A different approach is to consider that the network tries to maximize the utility of the users. This ambitious goal is thus a global optimization problem.

- One breakthrough was to realize that this global problem decomposes into subproblems that require only local information. Essentially, TCP approximates a solution to these subproblems by deciding how to drop packets in routers and how to control the window size in the hosts.

- A slightly different formulation results in backpressure protocols that solve jointly congestion control, routing, and scheduling. Thus, instead of patching up the protocols in the different layers by adding inter-layer headers and complex adjustments based on these headers, this approach enables us to design systematically protocols with provable optimality properties.

8.9 PROBLEMS

P8.1 Flows 1 and 2 use the network as shown in Figure 8.14. Flow 1 uses links AB and BC with capacities a and b, respectively. Flow 2 uses only link AB. Let $x = (x_1, x_2)$ where x_1 and x_2 denote the rates of the flows 1 and 2, respectively.

   (a) Find x that maximizes $x_1 + x_2$ subject to the capacity constraints.

   (b) Find x that maximizes $min(x_1, x_2)$ and satisfies the capacity constraints.

   (c) Find x that maximizes $\log(x_1) + \log(x_2)$ subject to the capacity constraints.

   (d) Write the gradient algorithm that solves the primal-dual problem of part (c).

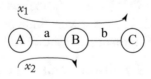

Figure 8.14: Figure for models Problem 1.

P8.2 There are N users, each with access to a different channel with each channel having bandwidth of W Hz. Assume that the transmission by user i encounters the noise power of $n_i > 0$ Watts at the receiving end. Furthermore, assume that user i with P_i Watts of power is able to achieve the data rate given by the Shannon Capacity formula.

   (a) Find the algorithm for allocating the total power of P Watts among N users such that the sum of the data rates achieved by the users is maximized.

(b) Suppose $P = 10$ Watts and $N = 4$. Find the optimal power allocation if the vector (n_1, n_2, n_3, n_4) for the noise powers for the users is $(3, 2, 1, 4)$ Watts. What would be the optimal allocation if this vector is $(3, 2, 1, 8)$ Watts?

P8.3 Consider the network shown in the Figure 8.15. External arrival is Poisson with the rate 100 packets/second, service times at the two queues are iid exponential with the rates of 200 and 100 packets/second as shown, and after service at Queue 1, each packet is independently sent to Queue 2 with probability 0.2, and leaves the network with probability 0.8. Find the average total delay through this network.

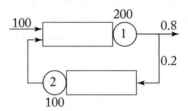

Figure 8.15: Figure for models Problem 3.

P8.4 A business site has two transmission lines to serve the packets generated from this site. The two lines are modeled by two servers and the corresponding queues as shown in the figure. The line represented by Server1 was the original one, and the second line was added at a later time to reduce the average delay experienced by the packets in getting out from this site. However, the second line is a slower line, and we need to determine the optimal probability p with which each arriving packet is independently directed to this slower line. Assume that the packets are generated from this site according to a Poisson process with rate of $\lambda = 1,000$ packets/second. For part (a) below, we also assume that transmission times for the two lines, modeled as the service times at the two servers, are exponentially distributed with the respective rates of $\mu_1 = 4,000$ and $\mu_2 = 3,000$ packets/second.

(a) Find the optimal value of p that minimizes the average delay experienced by the packets generated from this site.

(b) Now suppose the transmission times for the second line (the slower one) always have the fixed values of $1/3$ ms, while the ones at the first line are still exponentially distributed with the rate of $\mu_1 = 4,000$ packets/second. This could happen, for example due to different overheads for a function like encryption at the two lines, etc. How do you expect the optimal p for this scenario to compare (i.e., same, lower, or higher) to the one found in part (a) above? Justify your answer.

P8.5 f_1 and f_2 are two TCP connections in the Figure 8.17. The capacity of the bidirectional links AB and BC are 10 Mbps and 2 Mbps, respectively. For all questions below, ignore

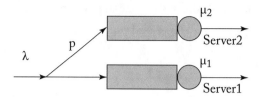

Figure 8.16: Figure for models Problem 4.

the effect of Fast Recovery (i.e., you can view the congestion avoidance phase as a simple AIMD with the multiplicative factor of 1/2).

(a) Suppose that only f_1 is active, and that f_1 is in the congestion avoidance phase. The average RTT for f_1 is 50 ms. Calculate the long-term average throughput of f_1.

(b) Now suppose that f_2 is also active and is in the congestion avoidance phase. The average RTTs for f_1 and f_2 are 50 ms and 100 ms, respectively. Calculate the long-term average throughput of f_1 and f_2.

(c) We want to maximize the total utility of this network. Using log utility function, write down an optimization problem and solve for the optimal rates. How does the result compare with that of part (b)?

(d) Write down a gradient algorithm in pseudocode that solves the above as a distributed congestion control problem.

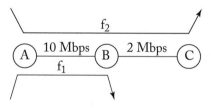

Figure 8.17: Figure for models Problem 5.

P8.6 Consider the network of two queues in series as shown in the Figure 8.18. Arrivals can be modeled by a Poisson process of rate x, and the two queues have service times that can be modeled as independent and identically distributed exponential random variables with the rates $\mu_1 = 12$ and $\mu_2 = 7$, respectively.

(a) Give an expression for total average delay across the network, denote it by $D(x)$.

(b) We want to maximize $f(x) = 10 * \log_e(x) - 2 * x$, subject to $D(x) \leq 0.4$ seconds. Give an expression for the Lagrangian for this problem.

(c) Show that the Slater conditions are satisfied for this problem.

(d) Find the optimal value for the arrival rate x.

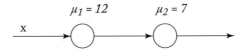

Figure 8.18: Figure for models Problem 6.

8.10 REFERENCES

The formulation of congestion control as a user utility maximization problem is due to Kelly et al. [54]. In that seminal paper, the authors show that the dual problem decomposes into a network problem and individual user problems. See also [61]. The concept of α-fairness was introduced in [71]. The theory of convex optimization is explained in [20]. The backpressure-based scheduling was introduced by Tassiulas and Ephremides in [100] and combined with congestion control by Neely et al. in [77]. For a detailed exposition of these ideas, see [95] and [96]. Libin Jiang added MAC optimization (see [51]).

CHAPTER 9

LTE

Cellular communication has revolutionized the way we communicate. It has enabled exchange of all forms of information (i.e., data, voice/audio, and video) from almost anywhere at anytime. Long-Term Evolution (LTE) is the most recent technology for implementing cellular networks. In this chapter, we first review the basic considerations needed in architecting a cellular network, and then explain the basics of an LTE network noting how it advances the capabilities of cellular networks. We also describe the key new technologies used in the evolved version of LTE referred to as LTE-Advanced. We end the chapter with a preview of the forthcoming 5G technology.

9.1 CELLULAR NETWORK

The goals of a cellular network are to provide global mobile access and good quality of data, voice, audio, and video transmissions. It attempts do so over a medium that is inherently unreliable due to artifacts like signal attenuation, blocking, and reflections. This lack of reliability is further exacerbated by user mobility, possibly at a high vehicular speed. Furthermore, user devices (e.g., cell phones, tablets, etc.) are typically power-limited in this environment. Users must be reachable wherever they are and the connections must be maintained when users move.

The key design idea of a cellular network is to divide a geographic area into *cells*. Users in a cell communicate with a *base station* (BS) at the center of the cell. Typically, neighboring cells use different frequencies to limit interference. Figure 9.1 sketches an idealized view of such an arrangement, where f_i denotes the central frequency used by a given cell. The interference is reduced by adapting the transmission power to the distance between the user devices and the BS. More recently, neighboring cells are able to use the same frequency using advanced interference mitigation techniques. For meeting the goal of reliable service, unlike WiFi, a cellular network operates in licensed frequency bands. Furthermore, data traffic between the user devices and a BS is tightly regulated by the BS to avoid contention and meet the performance requirements of the active connections. A BS is connected by a wired network to nodes performing the task of routing data traffic to and from the Internet and other phone networks, and is assisted by nodes performing functions required for establishment and maintenance of user connections. The overall architecture used for these additional nodes is an integral part of designing a cellular network.

To connect to the network, a user device listens to signals from BSs. It selects the BS with the most powerful signal. Each BS specifies periodic time slots during which new users can indicate their desire to connect. The user device sends a request during one of these time slots

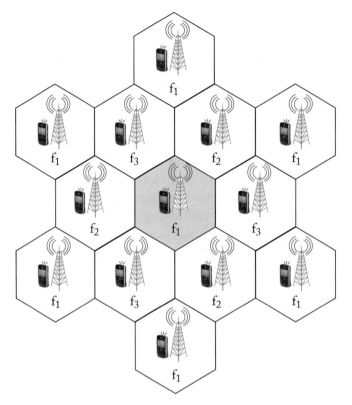

Figure 9.1: Frequency planning.

for the selected BS. If the request does not collide with one from another user, the BS allocates specific time slots and frequencies to communicate with the user device.

Each user is associated with a "home registry" that records her location. Periodically, a user device sends a signal that the local BS receives and then sends a message to the user's home registry to update her location. To contact a user, the network checks her home registry to find her location and then routes the information to the corresponding BS. When a user moves from one cell to another, the new BS takes over the communication in a process called *handover*. Consider a BS that is communicating with a user that is getting closer to a different BS. The user periodically reports to the network the signal strength from the two BSs, and the network helps the user undergo a handover to the new BS.

A given BS transmits at a total rate determined by the modulation scheme. This rate is shared by all the active users in the cell. Accordingly, the number of active users in the cell is bounded by the BS rate divided by the rate per user. Consequently, the area of the cell should be almost inversely proportional to the density of active users. (In fact, since the modulation scheme may depend on the quality of the user connection, the precise sharing is a bit more complex.) For instance, cells should be smaller in a dense metropolitan area than in a sparse rural area. Concretely, the radius of a cell ranges typically from 1–20 km. Note that when cells are small, handovers are likely to be more frequent. In cellular networks, the transmission power used is relatively small. For example, an LTE user device uses the maximum transmission power of about 0.2 Watts, while the maximum transmission power used by an LTE BS operating with a 10 MHz channel is about 40 Watts. Larger cells require relatively higher transmission power to and from the user devices.

In summary, a cellular network is required to provide high performance over a medium with unpredictable impairments while facing limitations on the frequency and power resources. In contrast to WiFi, a cellular network adds mobility support by tracking users and by performing handovers between BSs that are arranged to guarantee coverage across a wide geographic area.

After retracing the evolution to the LTE technology, we will discuss in the rest of this chapter how LTE addresses some of the tasks and challenges we outlined above.

9.2   TECHNOLOGY EVOLUTION

LTE is a specific cellular network implementation. It is standardized by the 3rd Generation Partnership Project (3GPP).

Figure 9.2 shows some of the important releases from 3GPP and the key technological milestones in evolution to LTE. The evolution has been a migration from circuit-switching to packet-switching, supporting mobility and multicast, and increasing the transmission rates.

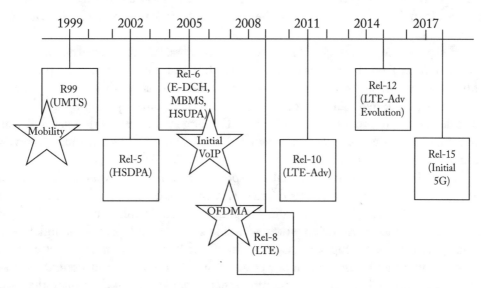

Figure 9.2: Technology evolution.

The release R99 specified Universal Mobile Telecommunication System (UMTS) that emphasized both mobile data and mobile voice services. UMTS supported voice using circuit switching and data using packet switching. Release 5 primarily targeted higher throughput and lower latency in the downlink direction by its High Speed Downlink Packet Access (HSDPA) specification. In a similar fashion, Release 6 targeted improved performance in the uplink direction by its High Speed Uplink Packet Access (HSUPA). This release made packetized voice using the IP protocol (Voice over IP - VoIP) feasible in practice with its support for the Quality of Service (QoS) required for real-time services. Additionally, this release introduced Multimedia Broadcast and Multicast Service (MBMS) for point-to-multipoint services. LTE, introduced with Release 8, made use of the physical layer based on the Orthogonal Frequency Division Multiplexing Access (OFDMA) technique; it is the main topic of this chapter. Releases 10 and 11 introduced advanced features for improving the performance of LTE. Toward the end of this chapter, we will briefly discuss these features in the LTE-Advanced section, and will also briefly discuss the forthcoming 5th Generation (5G) radio access technology expected to be addressed by 3GPP starting with its Release 15.

9.3   KEY ASPECTS OF LTE

The development of LTE is governed by a demanding set of requirements that includes downlink and uplink throughput of at least 100 and 50 Mbps, respectively, in a cell with 20 MHz of spectrum bandwidth, and data plane latency of under 5 ms across the Radio Access Network (RAN). Support for high speed mobility and significantly improved cell-edge performance are also required for LTE. LTE is required to be able to operate with channel bandwidth of 1.4–20 MHz.

LTE supports two techniques for the downstream and upstream transmissions to share the radio channels: Frequency Division Duplex (FDD) and Time Division Duplex (TDD). In the FDD mode, the upstream and downstream transmissions use different frequencies. In the TDD mode, these two transmissions share a common channel by dividing up its usage in time. The decision of whether to use FDD or TDD is largely dictated by the availability of the spectrum since FDD uses more spectrum than TDD. FDD is used in many more deployments than TDD.

LTE uses both open-loop and closed-loop power control schemes for uplink transmission to limit inter-cell interference and extend battery life in the user devices. For the uplink open-loop power control, a user device estimates the path loss of a downlink pilot signal of known strength, and adjusts its transmission power assuming similar path loss in the uplink direction. Additionally, for the closed-loop power control, the BS explicitly instructs a user device whether to increase or decrease its transmission power based on the received signal strength. The standards do not specify any downlink power control scheme, but leaves this decision to the network operator.

To maximize the total data rate that it offers in a reliable manner, LTE employs the following techniques: *Adaptive Modulation and Coding* (AMC), *Hybrid-Automatic Repeat reQuest* (H-ARQ), and *MIMO*.

AMC dynamically adapts modulation and coding used by a device in both downlink and uplink directions based on the prevailing channel condition as reported by feedback from the device. Since errors are more likely for modulation schemes that transmit bits faster, MAC uses such schemes only under good channel conditions.

In its Radio Link Control (RLC) sublayer, LTE implements the normal ARQ mechanism using Error Detection (ED) along with acknowledgments, timeouts, and retransmissions for reliable data delivery. Additionally, for a faster delivery of correct data, LTE also implements H-ARQ in its MAC sublayer. H-ARQ is a combination of an ARQ like scheme, Forward Error Correction (FEC), and Soft Combining as described below. LTE enables within its MAC protocol a quick indication of positive or negative acknowledgement for the data transmitted using H-ARQ. When one or more retransmissions are required, H-ARQ buffers all the previously received versions, combines them with the latest version, and then tries to decode the transmitted data. This is referred to as Soft Combining. In one method for Soft Combining, called *Chase Combining* (CC), the first transmission and all subsequent retransmissions contain the same data, ED overhead, and FEC overhead bits. CC combines the received copies using *Maximum Ratio Combining* (MRC). In MRC, before adding up the received signals, each version is multiplied by the complex conjugate of the channel gain for that reception. The other method for Soft Combining used by H-ARQ is called *Incremental Redundancy* (IR). The basic idea behind IR is to transmit more FEC related redundant information as additional transmission attempts are required. Since FEC can add significant overhead, IR attempts to save on the FEC overhead by transmitting it in steps on an as-needed basis. For doing so, IR prepares a small number of *Redundancy Versions* (RVs) in advance for each block of data, and after the first RV, the next RV is transmitted only if the receiver indicates that it is not able to decode the data from the previous RVs. Such transmission of the FEC overhead incrementally is made possible by use of the punctured Turbo codes. For example, the first RV may contain the data, ED overhead, and a small subset of the full FEC overhead, and the next RVs may contain just additional new subsets of the full FEC overhead. In the adaptive version of H-ARQ, depending on the channel condition, the sender can adjust with each attempt the transmission attributes like the subcarriers used for transmission, the RV transmitted, and the modulation and coding scheme used. If H-ARQ is unable to decode the transmitted data correctly in a limited number of retransmissions, it passes over the responsibility for receiving the correct data to the normal ARQ process.

As is the case for the more recent versions of WiFi (e.g., IEEE 802.11n), LTE supports MIMO for higher performance. In MIMO, an algorithm at the source determines what data each of its antennas transmits. For example, all antennas can transmit a copy of the same data for redundancy in order to improve reliability. This is referred to as transmit diversity, and is used when the channel conditions are poor. Another possibility is for each antenna to transmit different data to achieve a higher overall data rate. This is referred to as spatial multiplexing, and is used when the channel conditions are favorable. At the receiver, another algorithm combines the signals from the different receiving antennas.

9.3.1   LTE SYSTEM ARCHITECTURE

As shown in Figure 9.3, the BSs of an LTE network, called evolved Node Bs (eNBs), are interconnected by a wired IP network, called the Evolved Packet Core (EPC). This EPC consists of a *control plane* that carries control information and a *data plane* that transports the data to and from users. The data plane is attached to the Internet and to other phone networks. The EPC contains a Mobile Management Entity (MME), a Home Subscriber Server (HSS), a Serving Gateway (S-GW), a Packet Data Network Gateway (P-GW), and additional server(s) (not shown in the figure) for policy and charging rules. As shown in the figure, the S-GW transports uplink and downlink data traffic to and from the public or private Internet, respectively via the P-GW. The S-GW also plays the anchor point role for inter-eNB handover. The MME is dedicated to control plane functions, including user authentication for connection establishment, selection of an S-GW, handover assistance, tracking and paging of user devices in the idle mode, and security control. The figure shows control plane interactions between the MME and the eNB, and between the MME and the S-GW. For example, these interactions may contain signaling during the initial setup of a connection for user data flow, or signaling during a connection modification due to a handover. The MME also consults the HSS for obtaining the user subscription information, including user preferences. When a user is away from the home area, the user's designated HSS gets consulted for authentication as well as the user location information. An HHS includes the "home registry" function we discussed at the beginning of the chapter.

If a handover is within the same LTE system (i.e., intra-LTE handover), it is performed almost autonomously between the serving and target eNBs with minimal involvement of an MME and S-GW. Based on the channel quality reports from the user device for serving and neighboring eNBs, the serving eNB makes the handover decision and selects a target eNB. Subsequently, the serving eNB reserves the radio resources at the target eNB, coordinates the handover with the target eNB and the user device, and forwards the downlink user data to the target eNB until the new path from the S-GW to the target eNB is established. After establishing communication with the user, the target eNB informs the MME about the handover to start receiving the downlink user data directly via the S-GW. LTE also performs the other more complex forms of handovers: between two LTE systems (inter-LTE handover), and between LTE and another Radio Access Technology (RAT), e.g., a 3G technology (inter-RAT handover).

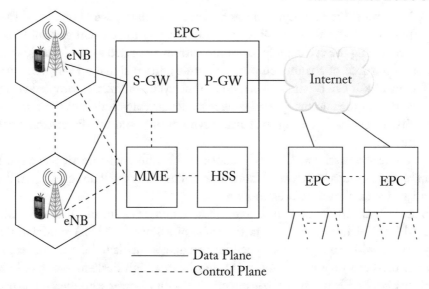

Figure 9.3: LTE system.

9.3.2 PHYSICAL LAYER

LTE uses a modulation technique called *Orthogonal Frequency Division Multiplexing* (OFDM). OFDM has a high spectrum efficiency. That is, OFDM is able to transmit a high data rate in a given spectrum bandwidth. To achieve this, OFDM transmits in parallel on a large number of closely placed subcarriers by modulating different data on different frequencies. To avoid these signals from interfering with each other, the modulation scheme is such that the maximum power of a subcarrier is at a frequency where the power of the other subcarriers is the smallest. This property holds if the subcarrier spacing is equal to the reciprocal of the symbol time, the duration over which a modulation symbol (i.e., the output of the modulation process) is transmitted.

To make the system robust against multi-path impairments that occur when the receiver gets a superposition of delayed versions of the transmitted signal along different paths, OFDM uses the following technique. Before transmission over each symbol time, OFDM prepends the signal representing the combination of the modulated subcarriers by a *cyclic prefix*, a portion of the tail of this combined signal. The fixed duration of this cyclic prefix is specified to be larger than the expected maximum delay spread, i.e., the maximum difference in the propagation times along different paths. In the discussion below we refer to the combined duration of the cyclic prefix and a symbol time as the extended symbol time. At the receiver, the cyclic prefix is discarded before decoding the signal. This provides robustness against inter-symbol interference as any delayed portion from the previous extended symbol time should fall within the duration of the cyclic prefix for the current extended symbol time. Furthermore, the cyclic prefix also provides protection against the intra-symbol interference. To see this we observe that the received signal (superposition of signals along different paths) during the current symbol time (i.e., the duration after removing the cyclic prefix from the current extended symbol time) is made up of superposition of portions of channel output of the transmitted signal for only the current symbol time. Technically, it can be shown that, due to the cyclic prefix, the sampled version of the received signal during the current symbol time is a circular convolution of the sampled version of the transmitted signal only during the current symbol time and the discrete channel impulse response function.

The multiple access scheme OFDMA allocates the different subcarriers and symbol times to data symbols (i.e., groups of data bits) from different users. We will discuss later in this section the granularity of such allocation used in LTE.

LTE uses OFDMA for downlink transmission, and a variant of OFDMA called Single Carrier–FDMA (SC-FDMA) for uplink transmission. Since OFDMA modulates each sub-carrier individually with data symbols, depending on data symbols to be transmitted, the transmission power used can vary significantly. This results in a high Peak-to-Average Power Ratio (PAPR). Since high PAPR translates to lower efficiency for a power amplifier, it is especially undesirable for generally power-limited user devices. SC-FDMA addresses this issue by inserting a preprocessing stage performing Discrete Fourier Transform (DFT) prior to subcarrier modulation. DFT produces linear combinations of the input data symbols, and hence reduces the variation in their magnitudes. This output from DFT is subsequently used for modulating the subcarriers. The presence of DFT at the transmitter requires an Inverse DFT (IDFT) stage at the receiver to retrieve the transmitted data symbols. Additionally, since DFT at the transmitter spreads information for a given data symbol across multiple subcarriers, channel equalization is required at the receiver to recover from any frequency selective fading. LTE performs this function digitally at the receiver by multiplying the received signal for each subcarrier by an appropriate complex number prior to (or along with the) IDFT operation. In spite of the benefit of lower PAPR, SC-FDMA is not deemed to be appropriate for the downlink direction since that would make the user devices more complex by necessitating the presence of the channel equalization function. The name SC-FDMA is used to highlight the fact that, due to the presence of DFT, this scheme shares some of the key characteristics of the traditional single carrier modulation scheme. In literature, SC-FDMA is also referred to as DFT-Spread FDMA.

Figure 9.4 shows the basic frame structure used for both downlink and uplink transmissions. One 10 ms Radio Frame is divided into 10 sub-frames, each of 1 ms duration. An eNB schedules uplink and downlink traffic on a sub-frame basis, i.e., every 1 ms. This is why a sub-frame is referred to as a *Transmission Time Interval* (TTI). A sub-frame is further divided into two 0.5 ms slots. Each slot accommodates the transmission times for 7 modulated symbols. A slot and 12 consecutive subcarriers form the smallest unit of allocation to a user referred to as a Resource Block.

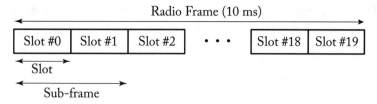

Figure 9.4: Frame structure.

Figure 9.5 illustrates a downlink Resource Block in the two-dimensional grid where symbol times are shown in the horizontal direction and subcarriers are shown in the vertical direction. Observe that a Resource Block consists of 7 symbol times (or one slot) by 12 subcarriers. Uplink Resource Blocks are defined in a similar fashion.

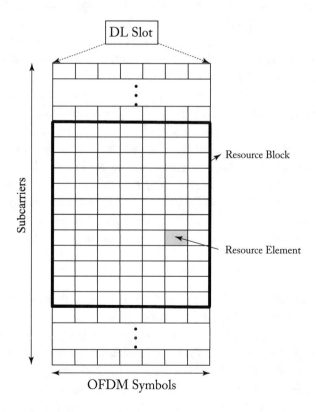

Figure 9.5: Resource grid.

In addition to the Resource Blocks for user allocations, LTE also defines a number of control channels for various functions related to connection management and system administration.

9.3.3   QOS SUPPORT

Traffic from different applications may have very different end-to-end performance requirements. For example, a voice or streaming video connection cannot tolerate a high delay, but can tolerate occasional packet loss, while transfer of a data file can usually tolerate somewhat higher delay, but with much smaller tolerance of packet loss. Hence, it is highly desirable that a cellular network is able to provide reasonable performance assurances for the traffic passing through it. We explain below the elaborate QoS framework LTE has incorporated in its specifications for achieving this goal.

When a user joins the LTE network, a default conduit (referred to as a default bearer) for transferring data of different user flows (e.g., for concurrent downloading of different online objects) across the LTE system is automatically set up with the QoS characteristics consistent with the user subscription. For data requiring special treatment, additional conduits (referred to as dedicated bearers) are set up. Each bearer is identified as either a Guaranteed Bit Rate (GBR) or a non-GBR bearer. A traffic flow is mapped to one of the bearers on the basis of the five-tuple used by its packets (source and destination IP addresses, source and destination transport layer port numbers, and the protocol used by the transport layer).

LTE provides class-based QoS support by associating a scalar Quality Class Identifier (QCI) with each bearer. QCI determines the packet treatment at each intervening node (e.g., scheduling weights, queue management thresholds, etc.). Note that LTE attempts to provide a consistent QoS treatment over the wireless leg as well as over the rest of the LTE system. Standardized QoS characteristics (e.g., whether the bearer is a GBR or a non-GBR bearer, tolerable levels of delay and packet loss rate, etc.) are associated with each QCI value. While QCI specifies the user plane treatment, Allocation and Retention Priority (ARP) specifies the control plane treatment for the purpose of either admitting a bearer or retaining a bearer due to a network event (e.g., loss of a link) requiring dropping of some of the established bearers.

9.3.4   SCHEDULER

The scheduler at an LTE BS (eNB) determines the user devices that get to receive and transmit in the cell supported by this eNB, and it provides differentiation opportunities among different service providers and vendors. Logically, there are separate downlink and uplink schedulers. The uplink scheduler announces allocations for uplink traffic based on their performance requirements and pending requests, while the downlink scheduler schedules downlink traffic based on their performance requirements and the data available for transmission at the eNB. These schedulers tend to be very complex as they attempt to do the optimal bin packing of the resource grid in the given direction while meeting the performance requirements of the underlying connections, and at the same time satisfying many scheduling constraints (e.g., requirements on contiguous allocation in frequency or time for certain traffic, timeliness requirements of many administrative and control traffic streams, deadlines for allocations for retransmitted traffic, etc.).

Across different classes of service, the schedulers commonly use some variant of prioritization or Weighted Fair Queueing (WFQ). For scheduling within a particular class, a scheduler called Proportionally Fair Scheduler (PFS) has been proposed. A PFS is often used as a reference scheduler for comparing other proprietary schedulers or new schemes from academic research. We discuss the basic idea behind the algorithm used by a PFS using a downlink scheduling scenario. Suppose n devices served by one eNB are downloading large volumes of data, and data for each devices is always available at the eNB for downlink scheduling.

The algorithm used by a PFS is a special case of the algorithm where device i is assigned a priority level that is proportional to $T_i^a(t)/R_i^b(t)$, where $T_i(t)$ and $R_i(t)$ are the instantaneous throughput achievable by device i in light of its current radio condition and the current estimate of the average throughput achieved by device i, respectively, and a and b are the parameters characterizing the scheduler. Average throughput for each device is estimated using an exponential smoothing algorithm based on the previous estimate and the recently realized instantaneous throughput. The device with the highest priority is scheduled first at each scheduling opportunity. With $a = 0$ and $b = 1$, the scheduler attempts to equalize the average throughput among the devices regardless of their current radio conditions, and can be seen as a round-robin scheduling scheme. Hence, if a partiuclar device happens to have a higher value for its $1/R_i(t)$, serving it first would increase its $R_i(t)$, and lower this device's priority. With $a = 1$ and $b = 0$, the scheduler attempts to maximize the instantaneous system throughput without taking into account any fairness consideration regarding the average throughput achieved by any device.

With $a = 1$ and $b = 1$, the scheduler attempts to strike a balance between these two extremes, and the associated scheduler is referred to as a PFS. The algorithm based on prioritization according to the $T_i(t)/R_i(t)$ value for each device turns out to be the solution of the problem that maximizes the sum of the logarithmic utilities of the devices. Here, the utility of device i is defined as $\log(\bar{R}_i)$ where $\bar{R}_i$ is the device's steady-state average throughput. As discussed in Section 8.4.1, the name PFS is justified due to this particular objective function. We state without proof that the solution of this problem, using the Primal-Dual theorem discussed in Section 8.4.2, shows that the optimal solution is achieved if the value $T_i(t)/R_i(t)$ for each device converges to an identical value. As in the case with $a = 0$ and $b = 1$, if a partiuclar device happens to have a higher value for its $T_i(t)/R_i(t)$, serving it first would increase its $R_i(t)$, and hence it would help equalize $T_i(t)/R_i(t)$ across the devices.

9.4   LTE-ADVANCED

The International Telecommunication Union (ITU) specified requirements in 2008 for a radio access technology to qualify to be designated as a 4th Generation (4G) technology. They included requirements that a mobile and a stationary device should be able to achieve peak downlink data rates of 100 Mbps and 1 Gbps, respectively. The baseline version of LTE discussed above did not meet these requirements. The baseline LTE version is commonly referred to as a 3.9G or a pre-4G technology. In this section, we briefly discuss the key advancements included in the later releases of LTE, referred to as LTE-Advanced. As suggested in Figure 9.2, these features were gradually introduced starting 3GPP's Release 10. LTE-Advanced was approved as a 4G technology by ITU in 2012.

For historic background, we note that 1G, 2G, and 3G radio access technologies are commonly associated with analog FDM based access, digital TDM based access and access defined by the UMTS architecture (where data communication used packet switching while voice communication used circuit switching), respectively. We will discuss the forthcoming 5G radio access technology in the next section.

9.4.1   CARRIER AGGREGATION

Carrier Aggregation allows an operator to offer a wider channel to its customers if the operator has license for more than one channel at a given location. Each carrier used for aggregation is called a component carrier, and can have bandwidth of 1.4, 3, 5, 10, or 20 MHz. Up to five component carriers can be combined to offer a wider channel of up to 100 MHz, and carrier aggregation for uplink and downlink are managed independently. These component carriers do not need to be contiguous and can even be in different bands. Since the component carriers may have different coverage due to the different signal attenuation properties, typically the component carrier with larger coverage is designated as the primary component carrier and additional secondary component carriers are added or removed with their availability at a given location. The primary component carrier is responsible for maintaining the radio connection with the user device.

9.4.2   ENHANCED MIMO SUPPORT

As compared to the support for 2x2 MIMO (i.e., 2 transmit and 2 receive antennas) in both uplink and downlink directions in LTE, LTE-Advanced supports up to 8x8 MIMO for the downlink communication and up to 4x4 MIMO for the uplink communication. In response to the changing radio environment, eNB controls the key communication attributes, like the number of transmit and receive antennas used, MIMO mode (transmit diversity vs. spatial-multiplexing), the number of independent data streams used by spatial-multiplexing, and the pre-processing scheme to be used. Different combinations of these attributed are standardized by defining different Transmission Modes (TMs). LTE-Advanced added new TMs to the set of TMs already supported by LTE.

9.4.3   RELAY NODES (RNS)

An RN is a low-power eNB that is connected to a normal eNB using a radio interface. The normal eNB in this arrangement is referred to as a Donor eNB (DeNB). While a DeNB is typically connected to an EPC using an optical fiber, an RN only uses radio interfaces on both user and network sides. RNs help extend the LTE coverage as well as enhance performance near cell edges. The radio resource of a DeNB is shared among the users directly supported by it and the associated RNs.The frequency spectrum used by an RN and its DeNB can be either the same or different. In the former case, the RN is referred to as an inband RN. In the inband scenario, network architects need to take into consideration the possibility of higher radio interference in determining the optimal placement of the RNs, and also consider the option of time-multiplexing of the frequency resources among the DeNB and the associated RNs.

9.4.4   COORDINATED MULTI POINT OPERATION (COMP)

The CoMP feature included in the newer LTE releases is also intended to improve performance at cell edges. This allows a user device to be simultaneously connected to more than one pair of transmit and receive points, where each pair of transmit and receive points is a set of co-located transmit and receive antennas at an eNB for the downlink and uplink communication, respectively. Different pairs of transmit and receive points used by CoMP for a given connection can be on different eNBs. For the downlink CoMP, data to be transmitted may be made available either at all or only one of the participating transmit points. In the former case, either Joint Transmission or Dynamic Point Selection can be used. Under Joint Transmission, a user device receives data from all participating transmit point simultaneously in each sub-frame. While under Dynamic Point Selection, a user device receives data from only one of the participating transmit points in a given sub-frame as dictated by the prevailing radio condition. On the other hand, if data to be transmitted is made available at only one of the participating transmit points, coordinated scheduling is used to determine the transmit point to be used for the given segment of data. Whether downlink data is made available at just one or all of the participating transmit points has a significant impact on the EPC traffic volume. For the uplink CoMP, the signal from a user device is received at multiple receive points and then combined for processing. The user device does not need to be aware of the presence of uplink CoMP. Joint Reception is possible under uplink CoMP, where the copies of the signal from a user device received at all the participating receive points over each sub-frame are used for joint processing. Clearly, for both uplink and downlink CoMP to work properly, tight coordination among the set of participating transmit/receive points is required for appropriate scheduling and/or data processing.

9.5   5G

The International Mobile Telecommunications for 2020 (IMT-2020) initiative of the ITU is in the process of defining the next-generation of the radio access technology commonly referred to as 5G. 3GPP has also embarked on the effort for next-generation radio access technology specifications to be submitted to IMT-2020 for a 5G designation. Release 15 of 3GPP is expected to include the initial set of 5G features.

The vision for IMT-2020 includes support for low-latency and high-reliability human-centric as well as machine-centric (i.e., machine-to-machine (M2M)) communication, high user density, high mobility, enhanced multimedia services, a wide range of needs related to energy consumption, performance and coverage, for Internet of Things (IoT) networking, and high precision positioning. As the core capabilities, IMT-2020 has identified the following eight parameters and their initial target values (which are to be investigated further). They include peak data rate (in Gbps), user data rate ubiquitously available in the target coverage area (in Mbps), radio network contribution to packet latency (in ms), maximum speed while supporting QoS and seamless handovers (in km/h), connection density (per km^2), network and device energy efficiency for reception and transmission (in bit/Joule), spectrum efficiency (in bit/s/Hz), and aggregate throughput per unit area within the target coverage (in Mbps/m^2). The respective initial target values are 20 Gbps, 10 Mbps, 1 ms, 500 km/h, 10^6 devices/km^2, 100x improvement over 4G, 3x improvement over 4G, and 10 Mbps/m^2. IMT-2020 is exploring feasibility of higher frequency spectrum in the 6–100 GHz for some of the deployment scenarios under its consideration. IMT-2020 also includes explicit support for the cutting-edge technologies like Software Defined Networking (SDN) and Network Function Virtualization (NFV) for lowering deployment and operational costs for the radio access network. We will discuss the topics of SDN and NFV in the Additional Topics chapter. Motivated by the same drivers as for SDN and NFV, IMT-2020 is also interested in the Cloud RAN (C-RAN) architecture that pools together at a centralized location resources for baseband and higher layer processing while connecting by fibers the radio heads and antennas deployed in the field as required for the intended coverage.

3GPP is targeting 5G deployments in two phases based on its Release 15 and 16, respectively. Each of these releases will include evolution of the RAN and overall system architecture. For RAN, 3GPP is considering both LTE-based enhancements as well as "New Radio (NR) for 5G." The 3GPP requirements for NR include similar requirements for the core capabilities considered by IMT-2020. At the present time, 3GPP is engaged in studying various competing proposals for NR.

9.6   SUMMARY

- Cellular networks rely on wireless access over a licensed portion of the spectrum. Their goal is to provide reliable and high-performance communication over an unreliable wireless channel while constrained by limited frequency and power resources.

- LTE is a broadband wireless access technology. It is the evolution of the technologies like UMTS, HSDPA, and HSUPA defined by the 3GPP, and is covered under its Release 8 specification. LTE has a simpler and flatter architecture as compared to that for the prior 3GPP technologies.

- LTE relies on technological advances like all IP communication, H-ARQ, MIMO, and OFDMA.

- LTE uses OFDMA for downlink transmission and SC-FDMA for uplink transmission. Both OFDMA and SC-FDMA allow time and frequency resources to be shared among different connections.

- The scheduler at an evolved Node B is the key network component that dictates the performance experienced by individual connections and also the overall efficiency with which the network resources are utilized. PFS is a reference scheduler that attempts to achieve a compromise between the overall system throughput maximization and strict round-robin fairness.

- LTE has extensive support for QoS. It aims to provide QoS support spanning the entire LTE system by using the concept of a bearer which is either GBR or non-GBR, and associating a QCI with each bearer.

- The technological advancements included in LTE-Advanced include Carrier Aggregation, enhanced MIMO support, Relay Nodes, and COMP.

- ITU's IMT-2020 initiative has published its vision for the 5G radio access technology that includes the targeted use cases and the anticipated core capabilities. 3GPP is investigating proposals for inclusion in its specifications which can qualify for the 5G designation.

9.7   PROBLEMS

P9.1 Consider a 10 MHz FDD LTE system with the following parameters for the DL transmission: 600 subcarriers for carrying data, and on average 25% of the capacity used up for signaling and connection management. Assume that all DL connections use QAM 64 modulation with coding rate of 1/2 (i.e., $6 \times 1/2 = 3$ data bits/subcarrier/symbol) and Single-Input Single-Output (SISO) operation.

(a) Find the maximum bit rate in DL that the network can possibly support.

(b) Assuming that a voice call uses 18 Kbps in each direction after accounting for all overheads, find the maximum number of simultaneous voice connections that the network can possibly support in the DL direction.

(c) Assume that a streaming video connection uses 400 Kbps only in the DL direction after accounting for all overheads. Plot the feasibility region of the number of voice calls and streaming video connections that can possibly be supported simultaneously in the DL direction.

P9.2 Consider a 10 MHz FDD LTE system with the following parameters for the DL transmission: 600 subcarriers for carrying data, and 25% of the capacity of each DL slot used up for signaling and connection management. Assume that all DL connections use QAM 16 modulation with coding rate of 3/4 (i.e., $4 \times 3/4 = 3$ data bits/subcarrier/symbol) and Single-Input Single-Output (SISO) operation. Suppose there are two 6 Mbps DL connections and two 3 Mbps DL connections. Each connection offers a data burst of constant size every 8 ms, where the size can vary across different connections. Find a scheduling scheme (i.e., how to map data in the DL Resource Grid) such that an entire burst from a given connection is accommodated in the same DL slot.

P9.3 In designing high data rate cellular networks, Doppler Shift consideration plays an important role. Doppler Shift can cause significant multipath interference for mobile users. Recall that the Doppler Shift or the change in the observed frequency can be estimated by $(\Delta v / c) f_0$, where Δv is the difference in the transmitter and receiver velocities, $c = 3 \times 10^5$ km/s is the speed of light, and f_0 is the emitted frequency.

  (a) For the channel with carrier frequency of 2.5 GHz, and maximum velocity (toward each other) between the transmitter and the receiver of 100 km/h, estimate the maximum Doppler Shift.

  (b) It can be shown that the channel Coherence Time (the duration for which the channel can be assumed to remain unchanged) $T_c \approx 1/\Delta_m f$ where $\Delta_m f$ is the absolute value of the maximum Doppler Shift. For the scenario of part (a), find the channel Coherence Time.

9.8   REFERENCES

The standards for LTE are described in [1, 3, 4, 5]. For further discussion on PFS, see [50, 58, 91, 96, 107]. The QoS control in LTE is discussed in [31]. SC-FDMA is explained in [75]. The basics of the new LTE-Advanced features are explained in [105]. The ongoing development of 5G is discussed in [87, 90]. The requirements for "New Radio" under consideration by 3GPP can be found in [2]. For a detailed discussion of wireless communication, including LTE, see [74]. [101] is a good reference for the fundamentals of wireless communication.

CHAPTER 10

QOS

The Internet provides a best-effort service. In principle, all the packets are treated equally. VoIP packets occasionally queue up behind email or file transfer packets. Packets from a video stream may have to be throttled to make room for a DVD download.

For many years, networking engineers have realized that the network could improve the quality of many applications by handling packets differently. Giving priority to urgent packets could have a minimal impact on applications that are not sensitive to delays. Some networks implement those ideas. This chapter discusses the general topic of quality of service.

10.1 OVERVIEW

Consider the set of flows that are sharing links in the Internet. These flows correspond to applications that have very different requirements that can be described by a minimum throughput R_{min} and a maximum delay D_{max}. Table 10.1 shows a few examples with approximate values and indicates the wide range of requirements of applications. These vast differences suggest that differentiated treatments of the packets would improve the user experience of different applications.

Table 10.1: Requirements of different applications

Application	R_{min} (Kbps)	D_{max} (s)
Video Streaming	60	10
VoIP	10	0.2
Download	30	200
Web Browsing	50	2
Video Conference	80	0.2

The important questions concern how and where packets should be treated differently. In Section 10.2 we explain traffic-shaping methods. In Section 10.3, we review scheduling mechanisms that can implement differentiated services. The end-to-end principle invites us to think about implementing the differentiated service in the end devices, without modifying the layers 1–3. We explore that possibility in Sections 10.5 and 10.6. Finally, in Section 10.7, we explore the economics of differentiated services, focusing on the issue of net neutrality.

10.2 TRAFFIC SHAPING

A router gets congested when packets arrive faster than it can transmit them. That situation certainly arises when the average arrival rate of packets is larger than the output link rate. It also occurs when packets arrive in large bursts. Leaky buckets are a simple mechanism to limit the average rate and the size of bursts.

10.2.1 LEAKY BUCKETS

Figure 10.1 illustrates a *leaky bucket* that shapes a flow of packets. The packets arrive at a Packet Buffer that stores them until they are allowed to leave. Tokens arrive at a *token counter* at the rate of a tokens per second. The counter counts the tokens but it saturates at B tokens. To transmit a packet with P bytes, there must be at least P tokens in the counter. The system then subtracts P from the counter and sends the packet.

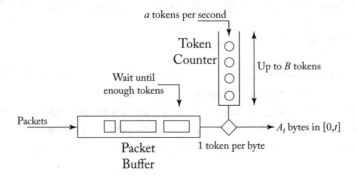

Figure 10.1: A leaky bucket shapes traffic by limiting its rate and burst size.

Let A_t designate the number of bytes that the system transmits in the interval of time $[0, t]$, for $t \geq 0$. From the description of the scheme, it should be clear that $A_{s+t} - A_s \leq B + at$ for $s, t \geq 0$. Indeed, the maximum number of tokens available to transmit packets in the interval $[s, s + t]$ is $B + at$. Thus, the long term average rate of A_t is bounded by a and the size of a burst is bounded by B. Note that for this scheme to work one requires the size of every packet to be at most B.

10.2.2 DELAY BOUNDS

If the rate and burst size of traffic are bounded, one can expect the delay through a buffer also to be bounded. This situation is shown in Figure 10.2 where N flows arrive at a buffer to be transmitted. Each flow n $(n = 1, \ldots, N)$ is shaped by a leaky bucket with parameters (a_n, B_n). The buffer can hold up to B bytes, and it transmits packets at the rate of C bytes per second. We assume that the size of every packet of flow n is bounded by B_n. Also, the buffer transmits only complete packets. We assume that at time 0 there are no packets in the system. We assume that

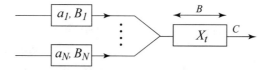

Figure 10.2: Buffer with shaped input traffic.

if a complete packet has entered the buffer at time t, then the system serves that packet before the packets that have not completely entered the buffer at that time. We also assume that each packet has at most P bytes and enters the buffer contiguously. That is, bytes of different packets are not interleaved on any input link. Let X_t indicate the number of bytes in the buffer at time t. One has the following result.

Theorem 10.1

Assume that $a_1 + \cdots + a_N \leq C$. Then
(a) One has

$$X_t \leq B_1 + \cdots + B_N + NP, t \geq 0.$$

(b) The delay of every packet is bounded by $(B_1 + \cdots + B_N + NP + P)/C$.

Proof. (a) Assume that $X_t > B_1 + \ldots + B_N + NP$ and let u be the last time before t that $X_u = NP$. During $[u, t]$, the system contains always at least one full packet and serves bytes constantly at rate C. Indeed, each packet has at most P bytes, so it is not possible to have only fragments of packets that add up to NP bytes. Because of the leaky buckets, at most $B_n + a_n(t - u)$ bytes of traffic enter the buffer in $[u, t]$ from input n. Thus,

$$X_t \leq X_u + \sum_{n=1}^{N}[B_n + a_n(t - u)] - C(t - u) \leq B_1 + \cdots + B_N + NP,$$

a contradiction.

(b) Consider a packet that completes its arrival into the buffer at time t. The maximum delay for that packet would arise if all the X_t bytes already in the buffer at time t were served before the packet. In that case, the delay of the packet before transmission would be X_t/C. The packet would then be completely transmitted at the latest after $(X_t + P)/C$. Combining this fact with the bound on X_t gives the result. □

10.3 SCHEDULING

Weighted Fair Queuing (WFQ) is a scheduling mechanism that controls the sharing of one link among packets of different classes. We explain that this mechanism provided delay guarantees to regulated flows. Both the definition of WFQ and its analysis are based on an idealized version of the scheme called *Generalized Processor Sharing* (GPS). We start by explaining GPS.

10.3.1 GPS

Figure 10.3 illustrates a GPS system. The packets are classified into K classes and wait in corresponding first-in-first-out queues until the router can transmit them. Each class k has a weight w_k. The scheduler serves the head of line packets at rates proportional to weight of their class. That is, the instantaneous service rate of class k is $w_k C/W$ where C is the line rate out of the router and W is the sum of the weights of the queues that are backlogged at time t. Note that this model is a mathematical fiction that is not implementable since the scheduler mixes bits from different packets and does not respect packet boundaries.

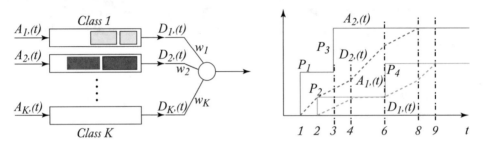

Figure 10.3: Generalized processor sharing.

We draw the timing diagram on the right of the figure assuming that only two classes (1 and 2) have packets and with $w_1 = w_2$. A packet P_1 of class 2 arrives at time $t = 1$ and is served at rate 1 until time $t = 2$ when packet P_2 of class 1 enters the queue. During the interval of time $[2, 4]$, the scheduler serves the residual bits of P_1 and the bits of P_2 with rate $1/2$ each.

From this definition of GPS, one sees that the scheduler serves class k at a rate that is always at least equal to $w_k C/(\sum_j w_j)$. This minimum rate occurs when all the classes are backlogged.

Theorem 10.2   *Assume that the traffic of class k is regulated with parameters (a_k, B_k) such that $\rho_k := w_k C/(\sum_j w_j) \geq a_k$. Then the backlog of class k never exceeds B_k and its queuing delay never exceeds B_k/ρ_k.*

Proof. The proof is similar to that of Theorem 10.1. Assume that the backlog X_t (in bytes) of class k exceeds B_k and let u be the last time before t that $X_u = 0$. During $[u, t]$, the scheduler serves at least $\rho_k [t - u]$ bytes of class k and at most $a_k (t - u) + B_k$ arrive. Consequently,

$$X_t \leq X_u + a_k (t - u) + B_k - \rho_k (t - u) \leq B_k$$

since $a_k \leq \rho_k$. This is a contradiction that shows that X_t can never exceed B_k.

For the delay, consider a bit of class k that enters the buffer at time t when the backlog of class k is X_t. Since the class k buffer is served at least at rate ρ_k, the time until that bit leaves cannot exceed $t + X_t/\rho_k$. Consequently, the delay cannot exceed B_k/ρ_k. $\square$

10.3.2 WFQ

As we mentioned, GPS is not implementable. Weighted Fair Queuing approximates GPS. WFQ is defined as follows. The packets are classified and queued as in GPS. The scheduler transmits one packet at a time, at the line rate. Whenever it completes a packet transmission, the scheduler starts transmitting the packet that GPS would complete transmitting first among the remaining packets. For instance, in the case of the figure, the WFQ scheduler transmits packet P_1 during $[1, 3.5]$, then starts transmitting packet P_2, the only other packet in the system. The transmission of P_2 completes at time 4.5. At that time, WFQ starts transmitting P_3, and so on.

The figure shows that the completion times of the packets $P_1, \ldots, P_4$ under GPS are $G_1 = 4.5, G_2 = 4, G_3 = 8, G_4 = 9$, respectively. You can check that the completion times of these four packets under WFQ are $F_1 = 3.5, F_2 = 4.5, F_3 = 7$, and $F_4 = 9$. Thus, in this example, the completion of packet P_2 is delayed by 0.5 under WFQ. It seems quite complicated to predict by how much a completion time is delayed. However, we have the following simple result.

Theorem 10.3   *Let F_k and G_k designate the completion times of packet P_k under WFQ and GPS, respectively, for $k \geq 1$. Assume that the transmission times of all the packets are at most equal to T. Then*

$$F_k \leq G_k + T, k \geq 1. \tag{10.1}$$

Proof.   Note that GPS and WFQ are work-conserving: they serve bits at the same rate whenever they have bits to serve. Consequently, the GPS and WFQ systems always contain the same total number of bits. It follows that they have the same busy periods (intervals of time when they are not empty). Consequently, it suffices to show the result for one busy period.

Let S_i be the arrival time of packet P_i.

Assume $F_1 < F_2 < \cdots < F_K$ correspond to packets within one given busy period that starts at time 0, say.

Pick any $k \in \{1, 2, \ldots, K\}$. If $G_n \leq G_k$ for $n = 1, 2, \ldots, k - 1$, then during the interval $[0, G_k]$, the GPS scheduler could serve the packets $P_1, \ldots, P_k$, so that G_k is larger than the sum of the transmission times of these packets under WFQ, which is F_k. Hence $G_k \geq F_k$, so that (10.1) holds.

Now assume that $G_n > G_k$ for some $1 \leq n \leq k - 1$ and let m be the largest such value of n, so that

$$G_n \leq G_k < G_m, \text{ for } m < n < k.$$

This implies that the packets $\mathcal{P} := \{P_{m+1}, P_{m+2}, \ldots, P_{k-1}\}$ must have arrived after the start of service $S_m = F_m - T_m$ of packet m, where T_m designates the transmission time of that packet. To see this, assume that one such packet, say P_n, arrives before S_m. Let G'_m and G'_n be the service times under GPS assuming no arrivals after time S_m. Since P_m and P_n get served in the same proportions until one of them leaves, it must be that $G'_n < G'_m$, so that P_m could not be scheduled before P_n at time S_m.

Hence, all the packets $\mathcal{P}$ arrive after time S_m and are served before P_k under GPS. Consequently, during the interval $[S_m, G_k]$, GPS serves the packets $\{P_{m+1}, P_{m+2}, \ldots, P_k\}$. This implies that the duration of that interval exceeds the sum of the transmission times of these packets, so that

$$G_k - (F_m - T_m) \geq T_{m+1} + T_{m+2} + \cdots + T_k,$$

and consequently,

$$G_k \geq F_m + T_{m+1} + \cdots + T_k - T_m = F_k - T_m,$$

which implies (10.1). □

10.4 REGULATED FLOWS AND WFQ

Consider a stream of packets regulated by a token bucket and that arrives at a WFQ scheduler, as shown in Figure 10.4.

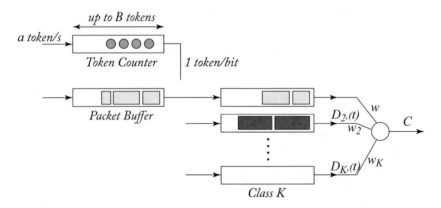

Figure 10.4: Regulated traffic and WFQ scheduler.

We have the following result.

Theorem 10.4   *Assume that $a < \rho := wC/W$ where W is the sum of the scheduler weights. Then the maximum queueing delay per packet is*

$$\frac{B}{\rho} + \frac{L}{C}$$

where L is the maximum number of bits in a packet.

Proof. This result is a direct consequence of Theorems 10.2–10.3. □

10.5 END-TO-END QOS

Is it possible to implement differentiated services without modifying the routers? Some researchers have suggested that this might be feasible.

Consider two applications, voice (telephone conversation) and data (downloading a web page with many pictures).

Say that the two applications use TCP. In the regular implementation of TCP, it may happen that voice gets a bit rate that is not sufficient. Consider the following setup.

First, we ask the routers to mark packets when their buffer gets half-full, instead of dropping packets when they get full. Moreover, the destination marks the ACKs that correspond to marked packets. (This scheme is called explicit congestion notification, ECN.) Normally, a TCP source should divide its window size by 2 when it gets a marked ACK. Assume instead that an application can maintain its normal window size but that the user has to pay whenever it gets a marked ACK. In this way, if the user does not care about the speed of a connection, it asks its TCP to slow down whenever it receives a marked ACK. However, if the user is willing to pay for the marks, it does not slow down.

The practical questions concern setting up an infrastructure where your ISP could send you the bill for your marks, how you could verify that the charges are legitimate, whether users would accept variable monthly bills, how to set up preset limits for these charges, the role of competition, and so on.

10.6 END-TO-END ADMISSION CONTROL

Imagine a number of users that try to use the Internet to place telephone calls (voice over IP). Say that these calls are not acceptable if the connection rate is less than 40 Kbps. Thus, whereas TCP throttles down the rate of the connections to avoid buffer overflows, this scheme might result in unacceptable telephone calls.

A simple end-to-end admission control scheme solves that problem. It works as follows. When a user places a telephone call, for the first second, the end devices monitor the number of packets that routers mark using ECN. If that number exceeds some threshold, the end devices abort the call. The scheme is such that calls that are not dropped are not disturbed by attempted calls since ECN marks packets early enough. Current VoIP applications do not use such a mechanism.

One slight objection to the scheme is that once calls are accepted, they might hog the network. One could modify the scheme to force a new admission phase every two minute or so.

10.7 NET NEUTRALITY

Network Neutrality is a contentious subject. Strict neutrality prescribes that the network must treat all the packets equally. That is, routers cannot classify packets or apply any type of differentiated service.

Advocates of an "open Internet" argue that neutrality is essential to prevent big corporations from controlling the services that Internet provides. They claim that a lack of neutrality would limit freedom of speech and the public good value of the Internet for users who cannot pay for better services. For instance, one can imagine that some ISPs that provide Internet access and phone services might want to disable or delay VoIP traffic to force the users to pay for the phone service. One can also imagine that some ISPs might provide preferential access to content providers with which they have special business relationships.

Opponents of neutrality observe that it prohibits service differentiation which is clearly beneficial for users. Moreover, being unable to charge for better services, the ISPs have a reduced incentive to upgrade their network. Similarly, content providers may not provide high-bandwidth content if the ISPs cannot deliver it reliably. It may be that most customers end up losing because of neutrality.

The dynamics of markets for Internet services are complex. No one can forecast the ultimate consequences of neutrality regulations. One reasonable approach might be to try to design regulations that enable service differentiation while guaranteeing that some fraction of the network capacity remains open and "neutral."

10.8 SUMMARY

The chapter discusses the quality of services in networks.

- Applications have widely different delay and throughput requirements. So, treating packets differently should improve some applications with little effect on others.

- One approach to guarantee delays to packet streams in the network is to shape the traffic with leaky buckets and to guarantee a minimum service rate with an appropriate scheduler such as WFQ.

- WFQ serves packets in the order they would depart under an idealized generalized processor sharing scheduler. We derive the delay properties of WFQ.

- An approach to end-to-end QoS would consist in charging for marked ACKs (either explicitly or implicitly by slowing down).

- An end-to-end admission control scheme can guarantee the quality of some connections with tight requirements.

- Should QoS be authorized? This question is the topic of network neutrality debates and proposed legislation.

10.9 PROBLEMS

P10.1 For the purpose of this problem, you can consider the idealized Generalized Processor Sharing (GPS) as being equivalent to Weighted Fair Queueing (WFQ).

Consider a system of four queues being serviced according to a WFQ scheduling policy as shown in the Figure 10.5. The weights given to the four queues (A, B, C, D) are 4, 1, 3, and 2, respectively. They are being serviced by a processor at the rate of 10 Mbps.

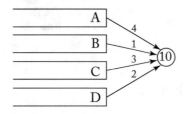

Figure 10.5: Figure for QoS Problem 1.

The Table 10.2 gives a list of different input traffic rates (in Mbps) at the four input queues. Fill in the resultant output rates for each of these four queues. We have filled in the first two rows to get you started!

Table 10.2: Table for QoS Problem 2

Input Rates				Output Rates			
A	B	C	D	A	B	C	D
1	1	1	1	1	1	1	1
10	10	10	10	4	1	3	2
6	6	2	2				
8	0	0	8				
1	5	3	5				

P10.2 Here we analyze the case of two applications sharing a 2 Mbps link as shown in the Figure 10.6. All packets are 1,000 bits long. Application A is video traffic which arrives at the link at a constant rate of 1 packet every millisecond. Application B is highly bursty. Its packets arrive as a burst of 10 Mbits worth of packets every 10 s. The incoming rate of the links carrying traffic from A and B is 10 Mbps each (this limits the peak rate at which packets enter the buffer for the link).

(a) The scheduler at the link is First Come First Serve (FCFS). If a packet of each of the applications arrives simultaneously, the packet of application A goes into the queue first (i.e., it is served first under FCFS). Calculate the following for Application A:

　i. The maximum packet delay.

　ii. Assume that this is the only link traversed by application A. What should the size of the playout buffer be at the receiver?

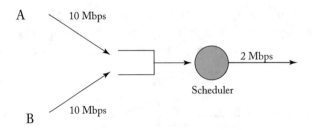

Figure 10.6: Figure for QoS Problem 2.

(b) Calculate the quantities above (in (a)i and (a)ii) for the case when the scheduler is round robin.

10.10 REFERENCES

The generalized processor sharing scheduling was invented by Kleinrock and Muntz [57] and Weighted Fair Queuing by Nagle [76]. The relationship between GPS and WFQ was derived by A. Parekh and Gallager [80] who also analyzed the delays of traffic regulated by leaky buckets. The possibility of bounding delays by regulating flows and using WFQ scheduling suggest a protocol where users request a minimum bandwidth along a path and promise to regulate their flows. Such protocols were developed under the name of RsVP (for reservation protocol) and IntServ (for integrated services). The readers should consult the RFCs for descriptions of these protocols. End-to-end admission control was proposed by Gibbens and Kelly in [36]. Wikipedia is an informative source for net neutrality and the debates surrounding that issue.

CHAPTER 11

Physical Layer

The most basic operation of a network is to transport bits. Communication links perform that task and form the *Physical Layer* of the network.

In this chapter, we describe the most common communication links that networks use and explain briefly how they work. Our goal is to provide a minimum understanding of the characteristics of these links. In particular, we discuss the following topics:

- Wired links such as DSL and Ethernet;

- Wireless links such as those of Wi-Fi networks; and

- Optical links.

11.1 HOW TO TRANSPORT BITS?

A *digital communication link* is a device that encodes bits into signals that propagate as electromagnetic waves and recovers the bits at the other end. (Acoustic waves are also used in specialized situations, such as under water.) A simple digital link consists of a flashlight that sends messages encoded with Morse code: a few short pulses of light—dashes and dots—encode each letter of the message and these light pulses propagate. (A similar scheme, known as the optical semaphore, was used in the late 18th and early 19th centuries and was the precursor to the electrical telegraph.) The light travels at about $c = 3 \times 10^8$ m/s, so the first pulse gets to a destination that is L meters away from the source after L/c s. A message that has B letters takes about B/R s to be transmitted, where R is the average number of letters that you can send per second. Accordingly, the message reaches the destination after $L/c + B/R$ s. The rate R depends on how fast you are with the flashlight switch; this rate R is independent of the other parameters of the system. There is a practical limit to how large R can be, due to the time it takes to manipulate the switch. There is also a theoretical limit: if one were to go too fast, it would become difficult to distinguish dashes, dots, and the gaps that separate them because the amount of light in each dot and dash would be too faint to distinguish from background light. That is, the system would become prone to errors if one were to increase R beyond a certain limit. Although one could improve the reliability by adding some redundancy to the message, this redundancy would tend to reduce R. Accordingly, one may suspect that there is some limit to how fast one can transmit reliably using this optical communication scheme.

Similar observations apply to every digital link: (1) The link is characterized by a transmission rate of R bits per second; (2) The signal propagates through the physical medium at some speed v that is typically a fraction of the speed of light c. Moreover, the link has some bit error rate BER equal to the fraction of bits that do not reach the destination correctly. Finally, the link has a speed limit: it cannot transmit bits reliably faster than its Shannon capacity C, even using the fanciest error correction scheme conceivable. (See Section 2.2.2.)

11.2 LINK CHARACTERISTICS

Figure 11.1 summarizes the best characteristics of three link technologies: wired (wire pairs or cable), wireless, and optical fibers. The characteristics shown are the maximum distance for a given bit rate, for practical systems. As the figure shows, optical fibers are capable of very large transmission rates over long distances. The Internet uses optical links at 10 Gbps (1 Gbps = 1 gigabit per second = 10^9 bits per second) over 80 km.

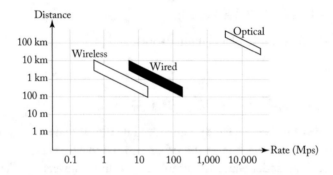

Figure 11.1: Approximate characteristics of links.

Wireless links such as used by WiFi achieve tens of Mbps over up to a hundred meters. A cellular phone links transmits at about 10 Mbps over a few kilometers.

Wired links of a fast Ethernet network transmits at 100 Mbps over up to 110 m. A DSL link can achieve a few Mbps over up to 5 km and about 10 Mbps over shorter distances. A cable link can transmit at about 10 Mbps over 1 km.

11.3 WIRED AND WIRELESS LINKS

A wireless link uses a radio transmitter that modulates the bits into electrical signals that an antenna radiates as electromagnetic waves. The receiver is equipped with an antenna that captures the electromagnetic waves. An amplifier increases the voltage of the received signals, and some circuitry detects the most likely bits that the transmitter sent based on the received signals. The signal must have a frequency high enough for the antenna to be able to radiate the wave.

A wired link, over coaxial cables or wire pairs, uses very similar principles. The difference is that the link can transmit much lower frequencies than a wireless link. For instance, the power grid carries signals at 60 Hz quite effectively. On the other hand, a wired link cannot transmit very high frequencies because of the skin effect in conductors: at high frequencies, the energy concentrates at the periphery of the conductor so that only a small section carries the current, which increases the effective resistivity and attenuates the signal quickly as it propagates. Another important difference between wired and wireless links is that wireless transmissions may interfere with one another if they are at the same frequency and are received by the same antenna. Links made up of non-twisted wire pairs interfere with each other if they are in close proximity, as in a bundle of telephone lines. Twisted pairs interfere less because electromagnetic waves induce opposite currents in alternate loops.

11.3.1 MODULATION SCHEMES: BPSK, QPSK, QAM

Consider a transmitter that sends a sine wave with frequency 100 MHz of the form $S_0(t) = \cos(2\pi f_0 t)$ for a short interval of T seconds when it sends a 0 and the signal $S_1(t) = -\cos(2\pi f_0 t)$ for T seconds when it sends a 1, where $f_0 = 100$ MHz. The receiver gets the signal and must decide whether it is more likely to be $S_0(t)$ or $S_1(t)$. The problem would be fairly simple if the received signal were not corrupted by noise. However, the signal that the receiver gets may be so noisy that it can be difficult to recognize the original signal.

The transmitter's antenna is effective at radiating electromagnetic waves only if their wavelength is approximately four times the length of the antenna (this is the case for a standard rod antenna). Recall that the wavelength of an electromagnetic wave with frequency f is equal to the speed of light c divided by f. For instance, for $f = 100$ MHz, one finds that the wavelength is 3×10^8 (m/s)$/10^8 s^{-1} = 3$ m. Thus, the antenna should have a length of about 0.75 m, or two and a half feet. A cell phone with an antenna that is ten times smaller should transmit sine waves with a frequency ten times larger, or about 1 GHz.

How does the receiver tell whether it received $S_0(t)$ or $S_1(t) = -S_0(t)$? The standard method is to multiply the received signal by a locally generated sine wave $L(t) = \cos(2\pi f_0 t)$ during the interval of T s and to see if the average value of the product is positive or negative. Indeed, $L(t) \times S_0(t) \geq 0$ whereas $L(t) \times S_1(t) \leq 0$. Thus, if the average value of the product is positive, it is more likely, even in the presence of noise, that the receiver was listening to $S_0(t)$ than to $S_1(t)$. It is essential for the locally generated sine wave $\cos(2\pi f_0 t)$ to have the same frequency f_0 as the transmitted signal. Moreover, the timing of that sinewave must agree with that of the received signal. To match the frequency, the receiver uses a special circuit called a phase-locked loop. To match the timing (the phase), the receiver uses the preamble in the bit stream that the transmitter sends at the start of a packet. This preamble is a known bit pattern so that the receiver can tell whether it is mistaking zeros for ones.

Using this approach, called *Binary Phase Shift Keying* (BPSK) and illustrated in Figure 11.2, the transmitter sends one bit (0 or 1) every T s. Another approach, called *Quadrature Phase Shift Keying* (QPSK), enables the transmitter to send two bits every T s. QPSK works as follows. During T s, the transmitter sends a signal $a \cos(2\pi f_0 t) + b \sin(2\pi f_0 t)$. The transmitter chooses the coefficients a and b among four possible pairs

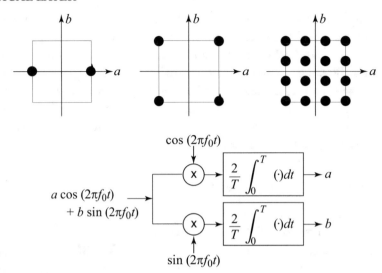

Figure 11.2: Top: BPSK (left), QPSK (middle), and QAM (right); bottom: demodulation.

of values $\{(-1,-1),(1,-1),(1,1),(-1,1)\}$ that correspond to the four possible pairs of bits $\{00, 01, 10, 11\}$ (see Figure 11.2). Thus, to send the bits 00, the transmitter sends the signal $-\cos(2\pi f_0 t) - \sin(2\pi t_0 t)$ for T s, and similarly for the other pairs of bits. To recover the coefficient a, the receiver multiplies the received signal by $2\cos(2\pi f_0 t)$ and computes the average value of the product during T s. This average value is equal to a. Indeed,

$$
\begin{aligned}
M(T) \quad &:= \int_0^T 2\cos(2\pi f_0 t)[a\cos(2\pi f_0 t) + b\sin(2\pi f_0 t)]dt \\
&= a\int_0^T 2\cos^2(2\pi f_0 t)dt + b\int_0^T 2\sin(2\pi f_0 t)\cos(2\pi f_0 t)]dt \\
&= a\int_0^T [1 + \cos(4\pi f_0 t)]dt + b\int_0^T [\sin(4\pi f_0 t)]dt.
\end{aligned}
$$

It follows that the average value of the product is given by

$$
\frac{1}{T}M(T) \approx a
$$

for T not too small. This demodulation procedure is shown in Figure 11.2. The value of the coefficient b is recovered in a similar way.

One can extend this idea to a larger set of coefficients (a, b). The quadrature amplitude modulation (QAM) scheme shown in Figure 11.2 chooses a set of 2^n coefficients (a, b) evenly spaced in $[-1, 1]^2$. Each choice is then transmitted for T s and corresponds to transmitting n bits.

Of course, as one increases n, the spacing between the coefficients gets smaller and the system becomes more susceptible to errors. Thus, there is a tradeoff between the transmission rate (n bits every T s) and the probability of error. The basic idea is that one should use a larger value of n when the ratio of the power of the received signal over the power of the noise is larger. For instance, in the DSL (digital subscriber loop) system, the set of frequencies that the telephone line can transmit is divided into small intervals. In each interval, the system uses QAM with a value of n that depends on the measured noise power.

11.3.2 INTER-CELL INTERFERENCE AND OFDM

In a cellular system, space is divided into cells. The users in one cell communicate via the base station of that cell. Assume that the operator has bought the license for a given range of frequencies, say from 1 GHz to 1.01 GHz. There are a number of ways that the operator can allocate that spectrum to different cells. One extreme way is to split the 10 MHz spectrum into disjoint subsets, say seven subsets of about 1.4 MHz each, and to allocate the subsets to the cells in a reuse pattern as shown in Figure 11.3. The advantage of this approach is that the cells that use

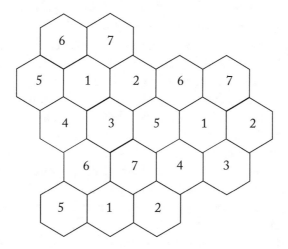

Figure 11.3: Cellular frequency reuse.

the same frequencies are far apart and do not interfere with each other. A disadvantage is that the spectrum allocated to each cell is small, which is wasteful if some cells are not very busy.

A different approach is to let all the cells use the full spectrum. However, to limit interference, one has to use a special scheme. We describe one such scheme: *Orthogonal Frequency Division Multiplexing* (OFDM). In OFDM, the spectrum is divided into many narrow subcarriers, as shown in Figure 11.4.

During each symbol time, the nodes modulate the subcarriers separately to transmit bits. A careful design of the subcarrier spacing and the symbol time duration makes the subcarriers

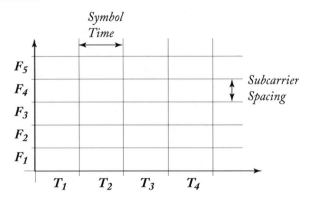

Figure 11.4: OFDM subcarriers and symbol times.

"orthogonal" to each other. This ensures that the subcarriers do not interfere with each other, and at the same time, for higher system efficiency, they are packed densely in the spectrum without any frequency guard bands. For example, an OFDM system with 10 MHz spectrum can be designed to have 1,000 subcarriers 10 KHz apart and the symbol time of 100 μs. The symbol times are long compared to the propagation time between the base station and the users in the cell. Consequently, even though some signals may be reflected and take a bit longer to reach a user than other signals, the energy of the signal in one symbol time hardly spills into a subsequent symbol time. That is, there is very little inter-symbol interference.

OFDM requires a single user to utilize the whole spectrum at any given time. It accommodates sharing between different users by time division multiplexing, i.e., allowing them to use the full spectrum during different symbol times. *Orthogonal Frequency Division Multiplexing Access* (OFDMA) is an evolution of OFDM where different users are allowed to share the spectrum at the same time. OFDMA uses subcarriers and symbol times to share the spectrum among the users by deploying both frequency and time division multiplexing. As discussed in the LTE chapter, LTE makes use of the OFDMA technology.

What about interference between different cells? The trick to limit such interference is to allocate the different subcarriers to users in a pseudo-random way in time and frequency with the latter being possible only if OFDMA is used. With such a scheme, one user interferes with another only a small fraction of the time, and the effect of such interference can be mitigated by using some error correction codes. One advantage is that if a cell is not very busy, it automatically creates less interference for its neighboring cells.

11.4 OPTICAL LINKS

The magic of optical fibers is that they can carry light pulses over enormous distances (say 100 km) with very little attenuation and little dispersion. Low attenuation means that the power of light after a long distance is still strong enough to detect the pulses. Low dispersion means that the pulses do not spread much as they travel down the fiber. Thus, pulses of light that are separated by gaps of a fraction of a nanosecond are still separated after a long distance.

11.4.1 OPERATION OF FIBER

The simplest optical fiber is called a step-index fiber, as shown in Figure 11.5. Such a fiber is made of a cylindrical core surrounded by a material with a lower refractive index. If the incident angle of light at the boundary between the two materials is shallow enough (as in the figure), the light beam is totally reflected by the boundary. Thus, light rays propagate along the fiber in a zig-zag path. Note that rays with different angles with the axis of the fiber propagate with different speeds along the fiber because their paths have different lengths. Consequently, some of the energy of light travels faster than the rest of the energy, which results in a dispersion of the pulses. This dispersion limits the rate at which one can generate pulses if one wishes to detect them after they go through the fiber. Since the dispersion increases linearly with the length of the fiber, one finds that the duration of pulses must be proportional to the length of the fiber. Equivalently, the rate at which one sends the pulses must decrease in inverse proportion with the length of the fiber.

The bottom part of Figure 11.5 shows a *single-mode fiber*. The geometry is similar to that of the step-index fiber. The difference is that the core of the single-mode fiber has a very small diameter (less than 8 microns). One can show that, quite remarkably, only the rays parallel to the axis of the fiber can propagate. The other rays that are even only slightly askew self-interfere and disappear. The net result is that all the energy in a light pulse now travels at the same speed and faces a negligible dispersion. Consequently, one can send much shorter pulses through a single-mode fiber than through a step-index fiber.

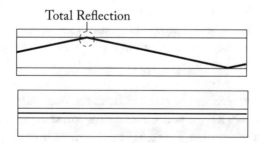

Figure 11.5: Step index fiber (top) and single-mode fiber (bottom).

11.4.2 OOK MODULATION

The basic scheme, called *On-Off Keying* (OOK) to send bits through an optical fiber is to turn a laser on and off to encode the bits 1 and 0. The receiver has a photodetector that measures the intensity of the light it receives. If that intensity is larger than some threshold, the receiver declares that it got a 1; otherwise, it declares a 0. The scheme is shown in Figure 11.6.

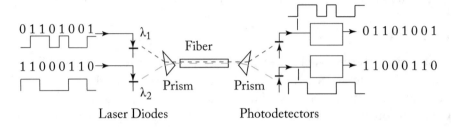

Figure 11.6: Wavelength division multiplexing. Different signals are encoded at different wavelengths.

To keep the receiver clock synchronized, the transmitter needs to send enough ones. To appreciate this fact, imagine that the transmitter sends a string of 100 zeroes by turning its laser off for 100 time T s, where T is the duration of every symbol. If the receiver clock is 1% faster than that of the transmitter, it will think that it got 101 zeroes. If its clock is 1% slower, it will only get 99 zeroes. To prevent such sensitivity to the clock speeds, actual schemes insert ones in the bit stream.

The minimum duration of a pulse of light should be such that the energy in the pulse is large enough to distinguish it from a background noise with a high probability. That is, even when in the absence of light, the receiver is subject to some noise that appears as if some light had impinged the photodetector. The source of this noise is the thermal noise in the receiver circuitry. Thus, in the absence of light, the receiver gets a current that is equivalent to receiving a random amount of light. In the presence of a light pulse, the receiver gets a random amount of light whose average value is larger than in the absence of a light pulse. The problem is to detect whether the received random amount of light has mean λ_0 or meant $\lambda_1 > \lambda_0$. This decision can be made with a small probability of error only if $\lambda_1 - \lambda_0$ is large enough. Since λ_1 decays exponentially with the length of the fiber as it gets divided by 2 every K kms, this condition places a limit on the length of the fiber for a specified probability of error. Note also that $\lambda_1 - \lambda_0$ is the average amount of light that reaches the receiver from the transmitter and that this quantity is proportional to the duration of a light pulse. This discussion suggests that the constraint on the length L of the fiber and the duration T of a light pulse must have the following form:

$$\alpha \times T \exp\{-\beta L\} \geq \gamma.$$

11.4.3 WAVELENGTH DIVISION MULTIPLEXING

Wavelength Division Multiplexing (WDM) is the transmission of multiple signals encoded at different wavelengths on the same optical fiber. The basic scheme is shown in Figure 11.6. Each bit stream is encoded as a signal using OOK. Each signal modulates a different laser. The light output of the lasers is sent through the fiber. At the output of the fiber, a prism separates the different wavelengths and sends them to distinct photodetectors. The photodetectors produce a current proportional to the received light signal. Some detection circuitry then reconstructs the bit stream.

Practical systems can have up to about 100 different wavelengths. The challenge in building such systems is to have lasers whose light wavelengths do not change with the temperature of the lasers. If each laser can be modulated with a 10 Gbps bit stream, such a system can carry 1 Tbps (= 10^{12} bps), for up to about 100 km.

11.4.4 OPTICAL SWITCHING

It is possible to switch optical signals without converting them to electrical signals. Figure 11.7 illustrates a micro electro-mechanical optical switch. This switch consists of small mirrors that can be steered so that the incident light can be guided to a suitable output fiber.

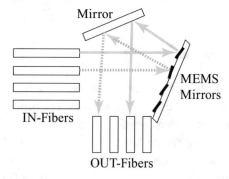

Figure 11.7: MEMS optical switch.

Such a switch modifies the paths that light rays follow in the network. The network can use such switches to recover after a fiber cut or some failure. It can also adjust the capacity of different links to adapt to changes in the traffic, say during the course of the day.

Researchers are exploring systems to optically switch bursts of packets. In such a scheme, the burst is preceded by a packet that indicates the suitable output port of the burst. The switch then has time to configure the switch before the burst arrives. The challenge in such systems is to deal with conflicts that arise when different incoming bursts want to go out on the same switch port. An electronic switch stores packets until the output port is free. In an optical switch, buffering is not easy and may require sending the burst in a fiber loop. Another approach is *hot-potato* routing, where the switch sends the burst to any available output ports, hoping that other switches will eventually route the burst to the correct destination. Packet erasure codes can also be used to make the network resilient to losses that such contentions create.

11.4.5 PASSIVE OPTICAL NETWORK

A *passive optical network* (PON) exploits the broadcast capability of optical systems to eliminate the need for switching. PONs are used as broadband access networks either directly to the user or to a neighborhood closet from which the distribution uses wire pairs.

Figure 11.8 illustrates a passive optical network. The box with the star is a passive optical splitter that sends the optical signals coming from its left to all the fibers to the different subscribers on the right. Conversely, the splitter merges all the signals that come from the sub-

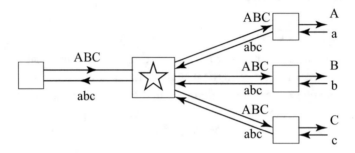

Figure 11.8: Passive optical network.

scribers into the fiber going to the distribution center on the left.

The signals of the different subscribers are separated in time. On the downlink, from the left to the right, the distribution system sends the signals A, B, C destined to the different subscribers in a time division multiplexed way. For the uplinks, the subscriber systems time their transmissions to make sure they do not collide when they get to the distribution center. The network uses special signals from the distribution center to time the transmissions of the subscriber systems.

11.5 SUMMARY

This chapter is a brief glimpse into the topic of digital communication.

- A digital communication link converts bits into signals that propagate as electromagnetic waves through some medium (free space, wires, cable, optical fiber) and converts the signals back into bits.

- The main characteristic of a link are its bit rate, length, and bit error rate. A wireless or wired link has a speed limit: its capacity, which depends on the range of frequencies it can transmit and on the signal-to-noise ratio at the receiver.

- Attenuation limits the length of an optical link, and dispersion limits the product of its length by the bit rate. WDM enables us to send different bit streams using different wavelengths.

- Optical switching is much simpler than electronic switching. However, because optical buffering is difficult, it requires new mechanisms to handle contention.

- The modulation schemes, such as BPSK, QPSK, and QAM, achieve a tradeoff between rate and error protection.

- A cellular network can use a frequency reuse pattern. OFDM coupled with interference mitigation techniques allows all the cells to use the full spectrum.

11.6   REFERENCES

Any book on digital communication covers the basic material of this section. See, e.g., [84]. For optical networks, see [86].

CHAPTER 12

Additional Topics

The previous chapters explained the main operation and organization principles of networks. In this chapter, we discuss a number of additional topics.

12.1 SWITCHES

The links of the Internet are connected by packet switches. The main function of a switch is to transfer packets from input links to output links with a minimum delay. Consider a switch with 32 input links and 32 output links. Assume that each link has a rate equal to 10 Gbps $= 10^{10}$ bps. Potentially, the total rate through the switch is then 320 Gbps. This rate exceeds the throughput of fast memory chips. Consequently, the switch cannot store all the arriving packets serially into memory and then read them out to the appropriate output links.

To go around this memory throughput limitation, engineers designed switch architectures where flows can proceed in parallel along different paths.

12.1.1 MODULAR SWITCHES

The left-hand part of Figure 12.1 illustrates a switch architecture called the Omega network for 8 input and 8 output links. The switch consists of 12 modules that are 2-by-2 switches. The

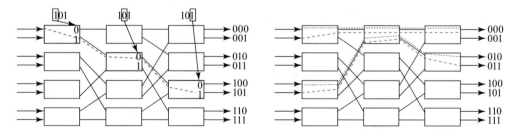

Figure 12.1: Omega switch (left) and hot spot (right).

figure also shows how the routing proceeds in each module. The first module chooses its output link based on the first bit of the output link, the second module based on the second bit, and the third module based on the third bit.

Assume that the flows are balanced in such a way that a fraction 1/8 of every input flow goes to each output link and that the arrival rates at every link is equal to R. In that case, the rate on every internal link of the switch is also equal to R. Hence, it suffices for each module to be able to handle a throughput equal to $2R$ for the overall switch to keep up with a throughput of $8R$.

If the flows are not balanced, hot spots may exist in the switch, as illustrated in the right-hand part of Figure 12.1. To prevent such a hot spot, one may use two Omega switches back-to-back and select the middle module randomly, as shown in Figure 12.2. The routing in the last three modules is as in Figure 12.1. The rate R in the figure is equal to $(R_0 + R_1 + R_4 + R_5)/4$. If all the links, external and internal, have the same rate, we see that R never exceeds that link rate. The same observation applies to all the links in the switch.

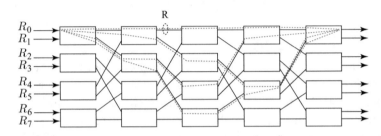

Figure 12.2: Balancing the flows.

The architecture in Figure 12.1 can be repeated recursively to build larger switches. For instance, if the modules are 8×8, one can use the architecture of Figure 12.1 where each line corresponds to 4 parallel lines. This results in a 32×32 switch.

If one does not insist that the switch can support the maximum possible rates, then other architectures can be used. Figure 12.3 illustrates a small example.

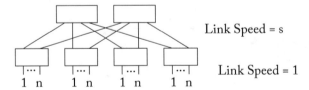

Figure 12.3: A fat tree network.

This regular arrangement forms a network with $4n$ inputs and outputs from smaller switches. In the figure, the links are bi-directional and have the indicated rates. Servers are attached at the bottom links. The rectangles represent packet switches.

In an arrangement common in data centers, 32, 48, or 64 servers are mounted on a rack, together with a switch called a *top-of-rack* switch. These switches are attached together via other switches, as in Figure 12.3. Thus, if each rack has 48 switches, the data center in the figure has 192 servers. Figure 12.4 shows a bigger arrangement, and it is easy to imagine a hierarchical assembly of such systems to interconnect a large number of servers. Note that the topology offers some redundancy. For instance, the leftmost bottom switch is not disconnected when one of its upper links fails. In addition to redundancy, the two links double the connection rate of that switch to the rest of the data center. The rate of the links that attach the servers to the top-of-rack switch are usually 10 Gbps and, in 2015, this rate is moving to 40 Gbps.

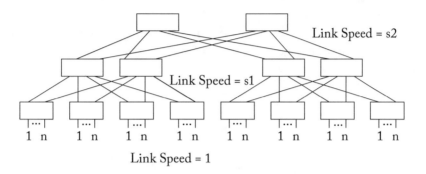

Figure 12.4: Another fat tree network.

It is typically the case that $2s < n$, so that the potential arrival rate n of packets from servers into a bottom-layer switch can exceed the maximum rate $2s$ that can go out of that switch via its upper links. Thus, if all these packets must go to an upper link to reach their destination, the switch can become saturated. The ratio $n/(2s)$ is called the *oversubscription ratio* of the bottom switches. For instance, if 48 switches are in a rack and attached with 10 Gbps, one may have s equal to 80 Gbps, by using two parallel links of 40 Gbs each. In this case, the oversubscription ratio is equal to 3. If that ratio is too large, the interconnection network may not work satisfactorily. In this particular network, the top-level switches are not over-subscribed: the total rate in cannot exceed the total rate out.

Assume that $n = 2s$ and that each switch can keep up with the packets as they arrive, with an acceptable delay. Then the network can deliver packets from any of the $4n$ servers to any other with an acceptable delay. However, this is the case only if the routing through the network is well-designed, as we discuss in the next section.

Another example, with three layers, is shown in Figure 12.4. It is also called a fat tree network. In this particular network, the bottom switches have an oversubscription ratio equal to $n/(s_1)$. The middle-switches may also be oversubscribed. Say that all the packets that arrive at their bottom links must go out through upper links. In that case, the maximum input rate is $4s_1$ and the maximum output rate is $2s_2$, which corresponds to a ratio $2s_1/s_2$. Assuming that $2s_1 > s_2$, such a middle switch can then be oversubscribed by that ratio.

12.1.2 SWITCHED CROSSBARS

Figure 12.5 shows another switch architecture called a switched crossbar. In this design, the input and output lines are attached to separate buffers connected by a crossbar switch. The switch operates in discrete time. In one time unit, the crossbar is configured to connect different inputs to different outputs. For instance, in the configuration shown in the figure, the connections are 4132, meaning that input 1 is connected to output 4, input 2 to output 1, and so on. During that time unit, the inputs send one packet in parallel to the outputs. The packets are of the same size and can be transferred in one time unit. When used for IP packets, the input buffers break them up into fixed-size chunks that the output buffers re-assemble.

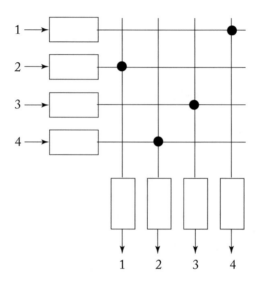

Figure 12.5: Switched crossbar.

Two questions arise in such an architecture: How should the buffers be arranged and how should the switch configure the crossbar at any given time?

To address the first question, assume that the input buffers are first-in-first-out and imagine that the packets at the head of the four input buffers are all destined for the same output. In that case, only one of the four packets can be transferred in the next time unit. This phenomenon, called *head-of-line blocking*, limits the throughput of the switch. Head-of-line blocking can be eliminated by maintaining four buffers for each input line: one for every possible output line. One calls this design *virtual output buffers*.

For the second question, assume that the switches uses virtual output buffers and say that the backlogs are $(3, 2, 0, 1), (0, 1, 3, 0), (0, 0, 1, 1), (2, 1, 2, 4)$. Our notation means that the first input line has 3 packets destined for output 1, 2 packets destined for output 2, and so on. Also, the fourth input line has 2 packets destined for output 1, 1 packet for output 2, and so on. In this situation, assuming that there is no new arrival of packets, the configuration shown in the figure would result in the following backlogs in the next step: $(3, 2, 0, 0), (0, 1, 3, 0), (0, 0, 0, 1), (2, 0, 2, 4)$. Is that a good choice? Note that there are $4! = 24$ possible configurations since each one corresponds to a permutation of the 4 outputs that are connected to the successive inputs. Which of these 24 configurations should the switch select?

A configuration is *maximal* if it transfers the maximum number of packets in the next step. Note that 4132 transfers only 3 packets since it connects input 2 to output 1 for which it has no packet. On the other hand, configuration 1234 transfers 4 packets, and so does configuration 4231. Thus, these last two configurations are maximal whereas 4132 is not. Intuitively, one might think that any maximal configuration is an equally good choice since it transfers the maximum number of packets in the next step. However, this view is myopic and it turns out that this choice does not maximize the throughput of the switch. The *maximum weighted configuration* achieves the maximum throughput. This is a configuration that maximizes the sum of the backlogs of the queues it serves in the next step, called the *weight of the configuration*. For instance, the weight of 4231 is $1 + 1 + 1 + 1 = 4$. The weight of 1324 is $3 + 3 + 0 + 4 = 10$ and is the maximum weight, even though this configuration is not maximal. Two facts are remarkable about the maximum throughput property of maximum weighted configurations. First, it does not depend on the arrival rates at the different links. Second, it is surprisingly simple: indeed, one might have expected that a good policy would depend in a very complex nonlinear way on the vectors of backlog and would not require simply comparing sums.

Unfortunately, even though this maximum weighted configuration is relatively simple, it is nevertheless too complex to be implemented in real time in a large switch: there are too many sums to compare. For that reason, switches implement much simpler round-robin policies that turn out to achieve a good throughput. One such policy is for the inputs to make requests to all the outputs for which they have packets, for the outputs to then grant the requests in a round-robin manner, and then for the inputs to accept the grant in a round-robin manner. That is, say that output 1 last granted a request of input 2. If in the next step inputs 3 and 4 make a request to output 1, the input 1 grants the request from input 3, which is next after the last grant 2 in the cyclic order $1, 2, 3, 4, 1, \ldots$. Similarly, say that input 1 last accepted a grant from output 3 and that in the next step it gets a grant from outputs $1, 2, 3$. It then accepts the grant from output 1. The switch may repeat this three-phase request-grant-accept process with the inputs that did not accept a grant yet. It is not possible to analyze the performance of this policy but simulations show that it performs well.

12.2 OVERLAY NETWORKS

By definition, an *overlay network* is a set of nodes that are connected by some network technologies and that cooperate to provide some services. In the Internet, the overlay nodes communicate via *tunnels*. Figure 12.6 illustrates such an overlay network. The four nodes A–D are attached to the network with nodes 1–12. In the figure, nodes u, x, y, z are other hosts.

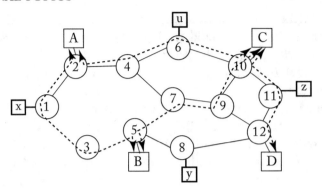

Figure 12.6: Overlay network: Nodes A, B, C, D form an overlay network.

Node A can send a packet to Node D by first sending it to node C and having node C forward the packet to node D. The procedure is for A to send a packet of the form [A | C | A | D | Packet]. Here, [A | C] designates the IP header for a packet that goes from A to C. Also, [A | D] are the overlay network source (A) and destination (D). When node C gets this packet, it removes the IP header and sees the packet [A | D | packet]. Node C then forwards the packet to D as [C | D | A | D | Packet] by adding the IP header [C | D]. We say that A sends packets to C by an IP tunnel. Note that the overlay nodes A–D implement their own routing algorithm among themselves. Thus, the network 1–12 decides how to route the tunnels between pairs of overlay nodes A–D, but the overlay nodes decide how a packet should go from A to D across the overlay network (A-C-D in our example). This additional layer of decision enables the overlay network to implement functions that the underlying network 1–12 cannot.

For instance, the overlay network can implement multicast. Say that x multicasts packets to u, y, z. One possibility is for x to send the packets to A, which sends them to C, that then replicates them and sends one copy to u, one to y, and one to z.

Another possibility is for the overlay network to perform performance-based routing. For instance, to send packets from y–u, node y can send the packets to B which monitors two paths to u: one that goes through A and one through C. The overlay network then selects the fastest path. One benefit is that such a scheme can react quickly to problems in the underlying network.

The overlay network can also provide storage or computing services.

12.2.1   APPLICATIONS: CDN AND P2P

We examine two specific examples: *Content Distribution Networks* (CDN) and *Peer-to-Peer Networks* (P2P Networks).

Content Distribution Network (CDN)

A widely used CDN is Akamai. It is an overlay network of servers. The idea is to serve users from a nearby CDN server to improve the response time of the system.

Consider once again the network in Figure 12.6. Assume that the nodes A–D are servers that store copies of some popular website. When node x wants access to the content of that website, it connects to the main server of that site that then redirects it to the closest content server (say A). Such redirection can happen in a number of ways. In one approach, the DNS server might be able to provide an IP address (of A) based on the source address of the DNS request (x). In an indirect approach, x contacts the main server of the website which then replies with a web page whose links correspond to server A.

Peer-to-Peer Network (P2P Network)

A P2P network is an overlay network of hosts. That is, end-user devices form a network typically for the purpose of exchanging files.

The first example of a P2P network was Napster (1999–2001). This network used a central directory that indicated which hosts had copies of different files. To retrieve a particular file, a user would contact the directory and get a list of possible hosts to download the file from. After legal challenges from music producers that Napster was an accomplice to theft, Napster was ordered to stop its operation.

Subsequent P2P networks did not use a central server. Instead, the P2P hosts participate in a distributed directory system. Each node has a list of neighbors. To find a file, a user asks his neighbors, who in turn ask their neighbors. When a host receives a request for a file it has, it informs the host that originated the request and does not propagate the request any further. Thus, the host that originates the request eventually gets a list of hosts that have the file and it can choose from which one to get the file. BitTorrent, a popular P2P network, arranges for the requesting host to download in parallel from up to five different hosts.

Once a host has downloaded a file, it is supposed to make it available for uploads by other hosts. In this way, popular files can be downloaded from many hosts, which makes the system automatically scalable.

12.2.2 ROUTING IN OVERLAY NETWORKS

The nodes of an overlay network typically have no knowledge of the details of the underlying network topology or capacities. This lack of information may result is serious inefficiencies in the routing across an overlay network.

12.3 HOW POPULAR P2P PROTOCOLS WORK

As of 2009, statistics (from www.ipoque.com) show roughly 70% of Internet traffic is from P2P networks and only 20% is from web browsing. Although there are big variations from continent to continent, 70% of the P2P traffic is using the BitTorrent protocol, and eDonkey contributes about 20% of the P2P traffic. Gnutella, Ares, and DirectConnect are the next popular P2P contributors. Although web-based multimedia streaming and file hosting appear to be gaining popularity, overall BitTorrent is still the king of the application layer traffic of the Internet!

Since the substantial amount of traffic is P2P related, a network engineer needs to understand how P2P protocols work in overlay networks. Proper understanding will help us design a better network architecture, traffic shaping policy, and pricing policy. In this section, we study how several popular P2P protocols work from the architectural point of view, but refrain from delineating the protocol specifications.

12.3.1 1ST GENERATION: SERVER-CLIENT BASED

UC Berkeley released a project in 1999 that enables distributed computing at networked home computers. The user side software performs time-frequency signal processing of extraterrestrial radio signals that is downloaded from a central SETI@HOME server, in the hope of detecting the existence of intelligent life outside the Earth. The SETI@HOME server distributes chunks of data to participating users to make use of their idle processing power, and the users return the computation results back to the server. The message path of SETI@HOME is of the Client-Server form. However, some classify this as the 1st generation of P2P. Strictly speaking, this is distributed grid computing.

12.3.2 2ND GENERATION: CENTRALIZED DIRECTORY BASED

The 2nd generation of P2P protocol is centralized directory based. Napster received a great deal of attention as an MP3 file-sharing service. Upon startup, a Napster client logs onto the central directory server and notifies the server of the list of its shared files. Clients maintain the connection with the server and use it as a control message path. When a client needs to search and download a file, it issues a query to the server. The server looks up its directory and returns the IP address of the owner of that file. Then the client initiates a TCP connection to the file owner and starts downloading. In this generation, the central server played a critical role of providing the search results. Therefore, it was easy to shut down the entire service by just shutting down the central server. A later version of Napster allowed client-client browsing.

12.3.3 3RD GENERATION: PURELY DISTRIBUTED

The primary feature of the 3rd generation P2P protocol is its fully distributed operation; it does not require the presence of a centralized server. Instead, every participating host works both as a client and a server. Gnutella operates in this manner, and has been quite successful. Before starting up, a client should have a list of candidate IP addresses. Upon startup, it tries to connect to them and learns the true list of working IP addresses. A search query is flooded over the Gnutella network over the active connections. A host that has the answer to the query responds along with its IP address. This scheme can suffer from scalability-related issues. The number of connections and the way of flooding can easily overwhelm the available bandwidth and the processing power. The later versions of Gnutella thus adopts the notion of super nodes.

12.3.4 ADVENT OF HIERARCHICAL OVERLAY—SUPER NODES

A P2P protocol in the pure form assumes that all hosts are equal. However, from the bandwidth and processing power considerations, some hosts are superior. By making them as relays for the network, the scalability issues can be mitigated. They are called super nodes in contrast to the normal leaf nodes. Introduction of the super nodes brings a hierarchy in the P2P overlay networks. Upon startup, a client connects to the super nodes instead of connecting to other clients directly. A search query is propagated via super nodes. Skype, a popular VoIP client, and the later version of Gnutella make use of the super nodes. Any Skype node with proper configuration can automatically be a super node. There have been reports that many university-based networked computers are serving as super nodes because of their higher speed Internet connections and openness.

There is the issue of being able to connect to the clients behind the NAT devices. If only the receiver client is behind a NAT device, the sender client is able to connect to it by issuing a call-back command. When both clients are behind the NAT devices, they use another Skype client residing in the public IP domain as a relay between the sender and the receiver.

12.3.5 ADVANCED DISTRIBUTED FILE SHARING: BITTORRENT

BitTorrent contributes a lion's share of the entire Internet traffic—it constitutes roughly 50% of it. Compared to other protocols explained above, the primary difference for BitTorrent lies in the greater redundancy in file distribution. A specialized web server, named tracker, maintains the list of the peers that are currently transferring a given file.[1] Any peer that wishes to download the file first connects to the tracker and obtains the list of active peers. A single file is split into many pieces, typically of 16KBytes, and is exchanged between the peers using these pieces. This enables a downloader to connect to multiple peers and transfer several pieces simultaneously. A unique policy called "rarest-first" is used to select the pieces. The downloader first downloads the least shared piece, increasing the redundancy of that piece over the network. After that, the downloader is able to serve as an uploader at the same time. So there is no pure downloader. All participating clients become peers. In order to discourage free riders who do not contribute to uploading, BitTorrent adopts the policy known as *Tit-for-Tat*, in which a peer provides uploading to the peers who reciprocally upload to it.

12.4 SENSOR NETWORKS

A *wireless sensor network* is a collection of sensor nodes that are equipped with a wireless transceiver. Depending on the application, the nodes can sense temperature, humidity, acceleration, intensity of light, sound, the presence of chemicals or biological agents, or other aspects of the environment. The nodes can relay packets to each other and to a gateway to the Internet or to some supervisor host.

---

[1]We do not differentiate between a Torrent Tracker and Indexer in our discussion.

Potential applications of this technology include environment monitoring to detect pollution of hazardous leaks, the health of a forest, the watering of a vineyard, the motion of animals for scientific investigations, the vital signs of patients, the structural integrity of a building after an earthquake, the noise around an urban environment, ground motions as indicators of seismic activities, avalanches, vehicular traffic on highways or city streets, and so on.

This technology has been under development for a dozen years. Although there are relatively few major deployments of wireless sensor networks to date, the potential of the technology warrants paying attention to it. In this section, we discuss some of the major issues faced by wireless sensor networks.

12.4.1   DESIGN ISSUES

The first observation is that which issues are important depend strongly on the specific application. It might be tempting to develop generic protocols for sensor networks as was done for the Internet, hoping that these protocols would be flexible enough to support most applications. However, experience has shown that this is a misguided approach. As the discussion below shows, it is critical to be aware of the specific features of the application on hand when developing the solution.

Energy

Imagine thousands of wireless sensor nodes deployed to measure traffic on a highway system. It is important for the batteries of these nodes to last a long time. For instance, say that there are 10,000 nodes with a battery life of 1,000 days. On average, 10 batteries must be replaced every day. Obviously, if the sensors have access to a power source, this issue does not matter; this is the case for sensors on board a vehicle or sensors that can be connected to the power grid.

Measurements of typical sensor nodes show that the radio system consumes much more than the other components of the nodes. Moreover, the radios consume about the same amount in the receive, idle, and transmit modes. Consequently, the only effective way to reduce the energy consumption of such nodes is to make the radio system sleep as much as possible. In the case of the highway sensors, it would be efficient to keep the sensors active for counting the traffic and turn on the radios periodically just long enough for the nodes to relay their observations. The nodes can synchronize their sleep/wakeup patterns by using the reception time of packets to resynchronize their clock.

Researchers have explored techniques for scavenging energy using solar cells, motion-activated piezoelectric materials, thermocouples, or other schemes.

Location

Imagine hundreds of sensor nodes dropped from a helicopter on the desert floor with the goal of detecting and tracking the motion of army vehicles. For the measurements to be useful, the nodes must be able to detect their location. One possibility is to equip each node with a GPS subsystem that identifies its precise location. However, the cost of this approach may be excessive.

Researchers have designed schemes for the nodes to measure the distances to their neighboring nodes. One method is to add an ultrasonic transceiver to each node. Say that node A sends a chirp and that its neighbor B replies with its own chirp. By measuring the delay, node A can estimate its distance to B. The chirps can include the identification of the chirping nodes. A similar approach is possible using radio transmitters, as used by airplane transponders.

The problem of determining locations from pairwise distances between nodes is nontrivial and has received considerable attention. The first observation is that if one builds a polygon with the sides of a given length, the object may not be rigid. For instance, a square can be deformed into a rhombus which has different node locations. Thus, a basic requirement to locate the nodes is that they must gather enough pairwise distances for the corresponding object to be rigid. A second observation is that a rigid object may have multiple versions that correspond to rotations, flips, or centrally symmetric modifications. The nodes need enough information to disambiguate such possibilities. A third observation is that even if the locations are unique given the measurements, finding an efficient algorithm to calculate these locations is also nontrivial and a number of clever schemes have been developed for this purpose.

A simpler version of the location problem arises in some applications. Say that every light fixture and every thermostat in a building has a unique identification number. However, the installer of these devices does not keep track of the specific location of each device. That is, the locations of all the nodes are known but the mapping of ID to location has to be determined after the fact. We let you think of viable approaches to solve this problem.

Addresses and Routing

Is it necessary for every sensor node to have a unique IP address? In the Internet, NAT devices enable us to reuse IP addresses. A similar approach can be used for sensor nodes. However, the nodes need to use UDP/IP protocols and simpler addressing schemes are conceivable. For instance, in some applications, the location may be a suitable address. If the goal is to find the temperature in a room of a building, the specific ID of the corresponding sensor does not matter. If the nodes are mobile, then a suitable addressing based on location is more challenging.

Routing in a sensor network may be quite challenging, depending on the application. If the nodes are fixed and always send information to a specific gateway, then the network can run some shortest path algorithm once or rarely. If the nodes are moving, the routing algorithm can either run in the background or nodes can discover routes when needed. As for the other protocols, routing should be designed with a good understanding of the characteristics of the application.

In-Network Processing and Queries

Say that the goal of a sensor network is to measure the highest temperature to which its nodes are exposed. One approach is for all the nodes to periodically report their temperature and for the supervising host to calculate the largest value. A different approach is for each node to compare its own temperature with the value that its neighbors report; the node then forwards only the maximum value. A similar scheme can be designed to measure the average value of the temperatures. More generally, one can design a communication scheme together with processing rules by the individual nodes to calculate a given function of the node measurements. The goal may be to minimize the number of messages that the nodes must transmit.

For a given sensor network, one may want to design a query language together with an automatic way of generating the processing rules and the messages to be exchanged to answer the query.

12.5   DISTRIBUTED APPLICATIONS

Networks execute distributed applications to implement many different protocols, including BGP, OSPF, RIP, and TCP. More generally, nodes on a network can be used to implement distributed applications for users. The properties of distributed applications are of interest to designers of the applications and the protocols. This section explores a number of representative applications and their properties.

12.5.1   BELLMAN-FORD ROUTING ALGORITHM

In the Routing chapter, we explained that the Bellman-Ford algorithm converges after a finite number of steps if the network topology does not change during that time. Recall that when running this algorithm, each node $i = 1, 2, \ldots, J$ maintains an estimate $x_n(i)$ of its shortest distance to a given destination node, say node D. Here, $n = 0, 1, \ldots$ denote the algorithm steps. In the basic version of the algorithm, if node i gets a new message $x_n(j)$ from a neighbor j, the node updates its estimate to $x_{n+1}(i)$ that it calculates as follows:

$$x_{n+1}(i) = \min\{x_n(i), d(i, j) + x_n(j)\}.$$

In this expression, $d(i, j)$ is the length of the link from i to j. The initial values are $x_0(i) = \infty$ for $i \neq D$ and $x_0(D) = 0$. If the network topology does not change, this algorithm results in non-increasing values of $x_n(i)$. Let the shortest path from i to D be the path $(i, i_1, i_2, \ldots, i_{k-1}, i_k, D)$. Eventually, there is one message from D to i_k, then one from i_k to i_{k-1}, and so on, then one from i_1 to i. After those messages, $x_n(i)$ is the shortest distance from i to D.

To make the algorithm converge when the topology changes, one has to modify the update rules. One modification is to let a node reset its estimate to ∞ if it gets a higher estimate from a neighbor. Such an increase indicates that the length of a link increased and that one should restart the algorithm. To implement this modification, the nodes must remember the estimates they got from their neighbors.

12.5.2   POWER ADJUSTMENT

Another distributed algorithm in networks is that used by wireless CDMA nodes to adjust their power. CDMA, for *code division multiple access*, is a mechanism that allows different nodes to transmit at the same time in some range of frequencies by assigning them different *codes* which makes their transmissions almost orthogonal. The details of this mechanism are not essential for our discussion. What is important for us is the observation that a transmission from one node can cause interference at the other nodes. The power adjustment problem is to find a scheme for the nodes to adjust their transmission power so that the communications between nodes are successful. The idea is that a transmitter may want to increase its power so that its receiver gets a more powerful signal. However, by increasing its power, the transmitter generates more interference for other receivers and deteriorates their operation. How can the nodes figure out a good tradeoff in a distributed way? This situation is not unlike TCP where an increase in the rate of one connection increases the losses of other connections.

To formulate the problem, imagine that there are pairs of (transmitter, receiver) nodes. Suppose for the pair i, the associated transmitter i sends packets to the associated receiver i with power P_i. Let $G(i, j)$ be the fraction of the power P_i that reaches receiver j. In addition, assume that receiver j also hears some noise with power η_j due to the thermal noise and the sources external to the nodes of the network under consideration. Thus, receiver j receives the power $P_j G(j, j)$ from the associated transmitter, noise power η_j, and interference power $\sum_{i \neq j} P_i G(i, j)$ from the other transmitters.

The key quantity that determines the quality of the operations of node j is the *signal-to-interference-plus-noise ratio, SINR*. For node j, this ratio is R_j, given by

$$R_j = \frac{P_j G(j, j)}{\sum_{i \neq j} P_i G(i, j) + \eta_j}.$$

That is, R_j measures the power of the signal $P_j G(j, j)$ from its transmitter divided by the sum of the power of the noise plus that of the interference from the other transmitters.

The communication from transmitter j to receiver j is successful if $R_j \geq \alpha_j$. Intuitively, if the signal is sufficiently powerful compared with the noise and interference, then the receiver is able to detect the bits in the signal with a high enough probability. Mathematically, the power adjustment problem is then to find the vector $P = (P_i)$ of *minimum* powers such that $R_j \geq \alpha_j$ for all j. We can write the constraint $R_j \geq \alpha_j$ as follows:

$$P_j G(j, j) \geq \alpha_j \left[\sum_{i \neq j} P_i G(i, j) + \eta_j \right].$$

Equivalently,

$$P_j \geq \sum_{i \neq j} P_i A(i, j) + \beta_j$$

where $A(i, j) = \alpha_j G(j, j)^{-1} G(i, j)$ for $i \neq j$ and $\beta_j = \alpha_j G(j, j)^{-1} \eta_j$. Defining $A(i, i) = 0$, we can write these inequalities in the vector form as

$$P \geq PA + \beta.$$

A simple adjustment scheme is then

$$P(n + 1) = P(n)A + \beta, n \geq 0$$

where $P(n)$ is the vector of powers that the transmitters use at step n of the algorithm. To explore the convergence of the algorithm, assume that

$$P^* = P^*A + \beta.$$

Then we find that

$$P(n + 1) - P^* = [P(n) - P^*]A,$$

so that, by induction,

$$P(n) - P^* = [P(0) - P^*]A^n, \quad \text{for } n \geq 0.$$

It can be shown that if the eigenvalues of matrix A have a magnitude less than 1, then $P(n)$ converges to P^*. Moreover, under that assumption it is easy to see that P^* exists and is unique. Let us now see how this algorithm works in a distributed manner. Consider the adjustment of power P_j. We find that

$$P_j(n + 1) = \sum_{i \neq j} P_i(n)A(i, j) + \beta_j = \alpha_j G(j, j)^{-1} \left[\sum_{i \neq j} P_i(n)G(i, j) + \eta_j \right].$$

We can write this update rule as follows:

$$P_j(n + 1) = \alpha_j G(j, j)^{-1} N_j(n)$$

where $N_j(n) = \sum_{i \neq j} P_i(n)G(i, j) + \eta_j$ is the total interference plus noise power that receiver j hears. Thus, if transmitter j knows $G(j, j)$ and $N_j(n)$, it adjusts its transmission power so that its receiver gets an SNIR exactly equal to the target α_j. To implement this algorithm, receiver j must indicate to its transmitter the value of $N_j(n)$ and also that of $P_j(n)G(j, j)$, so that the transmitter can determine $G(j, j)$.

Thus, the power adjustment of CDMA nodes has a simple distributed solution that requires a minimum amount of exchange of control information between the transmitters and receivers. The solution is simple because a receiver only needs to indicate to its transmitter the power of the signal it receives and the total power of interference and noise.

12.6   BYZANTINE AGREEMENT

Many applications in a network require nodes to coordinate their actions by exchanging messages. A fundamental question concerns the reliability of such coordination schemes when the network may fail to deliver messages. We consider some examples.

12.6.1 AGREEING OVER AN UNRELIABLE CHANNEL

Consider two generals (A and B) who want to agree whether to attack a common enemy to-morrow at noon. If they fail to attack jointly, the consequences are disastrous. They agree that general A will make the decision whether to attack or not and will then send a messenger with the decision to general B. General B can then send the messenger back to general A to confirm, and so on. However, the messenger has a small probability of getting captured when it travels between the generals. Interestingly, there is no protocol that guarantees success for generals A and B.

To appreciate the problem, say that general A sends a message with the decision to at-tack. General A cannot attack unless it knows that B got the message. To be sure, B sends the messenger back to A with an acknowledgment. However, B cannot attack until he knows that A got the acknowledgment. For that reason, A sends back the messenger to B to inform him that he got the acknowledgment. However, A cannot attack until he knows that B got his acknowledgment, and so on.

Let us prove formally that no algorithm can solve this agreement problem. By solving the problem, we mean that if general A decides not to attack, both A and B should eventually agree not to attack. Also, if A decides to attack, then A and B should eventually agree to attack. The proof is by contradiction. Say that some algorithm solves the problem. Consider the sequence of steps of this algorithm when general A decides to attack and all the messages get delivered. Assume that A and B agree to attack after n messages (the first one from A–B, the second one from B–A, and so on). Say that n is even, so that the last message is from B–A. Consider what happens if that message is not delivered. In that case, B still decides to attack. Thus, A knows that B will decide to attack whether A gets message n or not. Accordingly, A must decide to attack after it gets message $n - 2$. By symmetry, one concludes that B must decide to attack after it gets message $n - 3$. Continuing in this way, we conclude that A and B must agree to attack even if no message is exchanged. But then, they would also agree to attack even if A had decided not to attack, a contradiction. Hence, there cannot be an algorithm that solves the problem. A similar argument can be constructed for an odd n.

This problem has been analyzed when there is a lower bound on the probability that each message is lost and one can show that there is an algorithm that solves the problem with a high probability.

12.6.2 CONSENSUS IN THE PRESENCE OF ADVERSARIES

Consider that there are three agents A, B, C. One of them is dishonest but the other two do not know who he is. We want to design an algorithm where the honest agents agree on a common value in $\{0, 1\}$. In particular, if the two honest agents start with the same preferred choice $X \in \{0, 1\}$, then they should eventually agree on the choice X. It turns out that no such algorithm exists. More generally, in a network with N agents, no consensus protocol exists if there are at least $N/3$ dishonest agents.

We do not prove that no algorithm exists. Instead, we illustrate the difficulty by considering a sensible algorithm and showing that it does not solve the problem. In the first round of the algorithm, the honest agents report their choice and in the second round report what they heard from the other agents. We show that this algorithm does not solve the consensus problem. To show this, we consider three different executions of the algorithm and show that two honest agents may settle on different values (0 for one and 1 for the other), which violates the objective of the consensus algorithm.

Table 12.1 shows the case when A and B are honest and start with the choice 1 while C is dishonest and starts with the choice 0. The table shows the messages that the nodes send to each other in the two rounds. Note that in the second round C tells A that B told him that his choice was 0 (while in fact B reported his choice to be 1 in the first round).

Table 12.1: Case A(1), B(1), C(0) when C lies to A in the second round

→	A	B	C	A	B	C
A(1)	–	1	1	–	C = 0	B = 1
B(0)	1	–	1	C = 0	–	A = 1
C(0)	0	0	–	B = 0	A = 1	–

Table 12.2 considers the case when A is dishonest and starts with the choice 1 while B and C are honest and start with the choice 0. In the second round, A lies to C by reporting that B told him that his choice was 1 in the first round (instead of 0).

Table 12.2: Case A(1), B(0), C(0) when A lies to C in the second round

→	A	B	C	A	B	C
A(1)	–	1	1	–	C = 0	B = 1
B(0)	0	–	0	C = 0	–	A = 1
C(0)	0	0	–	B = 0	A = 1	–

Finally, Table 12.3 considers the case where B is dishonest and starts with the choice 1, while A and C are honest and start with the choices 1 and 0, respectively. In this case, B lies to C in the first round and all other messages are truthful.

To see why this algorithm cannot solve the consensus problem, note that in the first case A and B should end up agreeing with the choice 1, in the second case, B and C should agree with 0. Since A has the same information in the first and the third cases, he cannot distinguish between them. Similarly, C cannot distinguish between the second and the third cases. Since A must settle on 1 in the first case, he would settle on 1 in the third case as well. Similarly, since

Table 12.3: Case A(1), $\underline{B}$(1), C(0) when B lies to C in the first round

→	A	B	C	A	B	C
A(1)	-	1	1	-	C = 0	B = 1
$\underline{B}$(0)	1	-	$\underline{0}$	C = 0	-	A = 1
C(0)	0	0	-	B = 0	A = 1	-

C must settle on 0 in the second case, he would settle on 0 in the third case as well. Thus, in the third case, A and C end up settling on different values.

12.7 SOURCE COMPRESSION

Source compression is a method that reduces the number of bits required to encode a data file. One commonly used method is based on the *Lempel–Ziv algorithm*. To explain how the algorithm works, imagine that you want to compress a book. As you read the book, you make a list of the sentences that you find in the book, say of up to ten words. As you go along, you replace a sentence by a pointer to that sentence in your list. Eventually, it may be that you end up mostly with pointers. The idea is that the pointers take fewer bits than the sentences they point to.

The effectiveness of the scheme depends on the redundancy in the text. To appreciate this, consider using the same method for a binary file. Assume that you maintain a list of strings of 20 bits. If all the strings appear in the file, then you need 20-bit pointers to identify the strings. As a result, you replace a string of 20 bits by a 20-bit pointer, which does not achieve any compression. Now assume that there is quite a bit of redundancy so that you need only 8 bits to encode the 2^8 20-bit strings that show up in the file. In that case, each 20-bit string gets replaced by an 8-bit pointer, which achieves some significant compression.

This method is effective on binary strings corresponding to files with redundancy, such as text. For video and audio, different compression methods are used that exploit the specific features of such files.

12.8 SDN AND NFV

Software Defined Networking (SDN) and Network Function Virtualization (NFV) are two recent technologies that have drawn a lot of interest from both academia and industry. Expense control and operational flexibility are the two key driving forces behind these technologies.

SDN separates out the network data plane where the end-user data is forwarded and the network control plane where the rules for the data forwarding are determined. In this paradigm, the network data plane uses simple hardware boxes and links for forwarding end-user data according to the instructions they receive externally. Such simple hardware boxes are referred to as the "bare-metal boxes." On the other hand, the network control plane in this paradigm is "centralized." A network controller node communicates with the "bare-metal boxes" to stay current with its view of the network topology, and provides these boxes data-forwarding rules that have been derived by the algorithms running in applications tethered to the controller. The simplicity of the network data plane and the "centralized" control plane reduce both capital and operational expenses. Initially, data centers, e.g., inter-connection of a large number of servers at a Google data center, were the prime target. Subsequently, service providers like AT&T and Verizon also got interested in evolving their networks using the SDN technology. As the size of the network to be controlled by the SDN controller gets large, it is too demanding for a single controller node to control all the nodes in the network. This leads to the ideas of a set of distributed controllers, each responsible for a subset of the network nodes, but staying up-to-date with each other regarding their view of the network and still acting in a unified fashion for controlling the network, and thus giving the perception of a "centralized" controller. Currently, there are a number of SDN initiatives at various stages of implementation, some public and some proprietary. They all are built on the key concepts we discussed above. We will further explore the SDN-based architecture later in this section.

NFV implements a wide range of Network Functions (NFs) like packet forwarding, firewall, intrusion detection, etc., on general-purpose servers which may be located in the "cloud," i.e., at a data center hosting a large pool of servers. In the context of the service provider networks, the term "middlebox" is often used to refer to a dedicated network device that implements one such NF. Virtualized NFs (VNFs) require considerations like how many instances of an NF to be created (both initially and dynamically), which servers to use for these instances, how to perform load balancing among these instances, how to derive fault-tolerance, how to monitor and ensure their performance, etc. It is useful to have a common framework that addresses such common issues. We will discuss below one such framework called E2, a moniker derived from "Elastic Edge." A more evolved view of SDN is to think of it as the means of not only controlling packet-forwarding nodes, but all the VNFs as well.

12.8.1 SDN ARCHITECTURE

Figure 12.7 shows the basic SDN architecture with a single controller. As shown, the SDN controller communicates with the SDN applications and the network elements over its northbound and southbound interfaces, respectively. Using the southbound interfaces, the controller obtains the network topology and its current state. The SDN applications implement algorithms to come up with the recommendations for the network (e.g., routing path) based on the information received from the controller (e.g., the network topology), and relay them back to the controller. The controller, in turn, sends the pertinent information from these recommendations to the network elements over the southbound interfaces (e.g., for updating their routing tables). OpenFlow is an example of a protocol suitable for the southbound interfaces. We can now see that the SDN paradigm farms out the intelligent decision making to the SDN applications from the conventional network elements. This is the reason for the promise of reduction in both capital and operational expenses. It is also reasonable to expect that the relative difficulty in changing the way a network behaves would be lower with the SDN paradigm as that can be accomplished simply by changing the algorithms incorporated in the SDN applications.

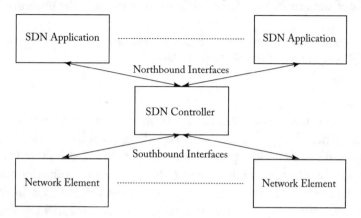

Figure 12.7: SDN architecture.

12.8.2 NEW SERVICES ENABLED BY SDN

SDN allows creation of new services without substantial changes in the network architecture. Examples of such innovative services include supporting connection with the prescribed performance guarantees, reliability, and/or security. To illustrate the basic idea, let us consider performance guarantees in terms of rate and the packet transfer delay limit. One of the ways of supporting such a service within the SDN framework is as follows. When a connection request arrives at an access node, the SDN controller is consulted for finding a route that can satisfy the requested performance guarantees. With help from an SDN application, the controller finds and returns an identifying tag to be used in the header of each packet of the connection, and also configures the nodes along the intended route to achieve the desired routing. For determining the route, the SDN application can take into account the utilization level of each intervening link. The link utilization levels can be obtained either by periodic measurements or by calculations after each route assignment, where the latter would need an activity time-based detection of continuation of a connection at its ingress node. We also note that performance guarantees would require the ingress node to make sure that the connection is not offering traffic in access of the requested rate, which can be achieved by simple Leaky Bucket-based monitoring. An alternate strategy can be based on implementation of multiple priority queues at each outgoing port of each node, and assigning each packet via its identifying tag to the queue at the appropriate priority level. For enhanced reliability, a similar approach can be used where the SDN controller instructs the network to replicate the packets of the given connection at the ingress node, route them along along two independent paths, and finally remove one of the copies of each packet at the egress node if both reach there. For enhanced security, the ingress and egress nodes can be provided encryption and decryption keys for a given connection, respectively by the SDN controller, and transporting encrypted packets across the network, and thus protecting the encapsulated data from any breach while moving through the unsecured portion of the network.

Configuration and management of the cloud resources used for NFV can also be thought of as the potential application of the SDN paradigm.

Deferrable Service

We illustrate below how new services requiring significant analysis, measurements, and control are made possible by the SDN paradigm. We follow the approach presented in [68].

Networks are dimensioned to provide services with small delays and high throughput during peak periods. Due to the sizable difference in the network utilization between the peak and off-peak periods as well as the requirements of robust performance in the face of both traffic burstiness and various types of network failures, these networks are significantly overdimensioned for the average network loads. This extra capacity can be used for supporting a new traffic class that can tolerate higher end-to-end delays with some guarantees. We refer to this class of traffic as deferrable traffic. The network ingress rate for deferrable traffic can be controlled in real time using the SDN framework. The SDN controller can obtain the current link-by-link utilization in the network by the normal non-deferrable traffic. These measurements can be used by an SDN application to estimate how much of deferrable traffic can be admitted in the network while still satisfying the corresponding delay guarantees with a high probability. We discuss below one possible way for deriving such estimates. We assume in our discussion that the non-deferrable traffic is treated at a higher priority over each link to make its performance essentially transparent to the presence of the deferrable traffic.

Let us consider a single backbone link between two nodes. Assume that each connection has identical throughput as constrained by the corresponding access link rate (which is assumed to be identical for all connections for this discussion), and the capacity C of the backbone link is expressed in the units of the number of connections it can support. We consider providing deferrable service at a normalized throughput of D, i.e., the throughput normalized by the common access link rate. Given the target probability of the delay seen by the deferrable traffic staying below a specified value of T, how large a value of D can the network allow? Observe that the deferrable traffic will have delay less than T if the total deferrable traffic served during T is at least DT. Here we assume that the deferrable traffic is served on a first-come-first-served basis after serving the non-deferrable traffic. Denoting the number of active non-deferrable connections by X_τ, we can express the target probability as follows.

$$P\left(\int_t^{t+T} (C - X_\tau)d\tau \geq DT\right) = 1 - P\left(\frac{1}{T}\int_t^{t+T} X_\tau d\tau > C - D\right). \tag{12.1}$$

The SDN controller will limit the amount of deferrable traffic such that this probability is acceptably large, say 99%. The situation is illustrated in Figure 12.8. In order to estimate the

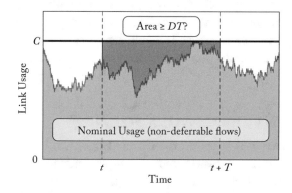

Figure 12.8: Leftover capacity for deferrable service.

probability of interest, it is convenient to first estimate the "failure" probability

$$P_{A,T} := P\left(\frac{1}{T}\int_0^T X_\tau d\tau > A\right). \tag{12.2}$$

Using the Large Deviations Theory, we can deduce

$$P_{A,T} = e^{-TI(A)+o(T)} \tag{12.3}$$

where the function $I(\cdot)$ is called the Large Deviations Rate function. As discussed in [68], using the measurement-based statistics of X_τ, it is possible to estimate this "failure" probability, and subsequently determine how large D can be while being transparent to the non-deferrable traffic and at the same time satisfying the delay requirements of the deferrable traffic. [68] also shows that the basic approach explained above can be extended to a general network beyond just a single link network considered above. This paper also analyzes the scenario where intermediate nodes on a route for deferrable connection can cache its traffic to wait for the opportune time to forward it over the next hop.

12.8.3 KNOWLEDGE-DEFINED NETWORKING

The SDN architecture shown in Figure 12.7 provides a natural platform for incorporating decision making based on Machine Learning. This can be done by invoking SDN applications that are learning capable. Such applications can undergo supervised learning, i.e., using past data for tuning a model that can estimate the variable(s) of interest. The past network measurements usually provide the data needed for this supervised learning. The term Knowledge-Defined Networking (KDN) is used by [22] to refer to the SDN architecture equipped with learning-capable applications. Linear Regression, Neural Networks, and Support Vector Machines are examples of the techniques for supervised learning. One simple example is to predict the traffic load during the next hour using the final model following the learning phase. One can use the training examples with similar time-of-day, day-of-week, and week-of-year hours. If one is able to predict traffic load with reasonable accuracy, it would be possible to reconfigure the network in a proactive manner. [7] has a collection of nice lectures with detailed explanations of the key Machine Learning techniques.

12.8.4 MANAGEMENT FRAMEWORK FOR NFV

Our discussion in this section is based on the Elastic Edge (E2) initiative from a Berkeley research group (see [24]). For the network functions implemented on generic servers using software constructs like Virtual Machines or Containers, there are common tasks required for managing each instantiation of the Virtualized Network Functions (VNFs). The management tasks include placement (on which server a given VNF instant runs), scaling (strategy for scaling the software resources allocated for a VNF as its load varies), composition (the order of the VNFs for a given class of traffic), failover (strategy for making the VNFs fault-tolerant), and QoS control (strategy for ensuring that each VNF instant performs per the anticipated Service Level Agreement (SLA)). Of course, it is possible to implement such management tasks in a customized manner for each VNF. However, that would be burdensome for the developers of the VNFs, possibly cross-interfering (with those for other VNFs), and possibly riddled with inefficient implementation. Hence, it is very desirable to have a common framework that can attend to all the management tasks for all VNF instants. Loosely speaking, this is akin to having an overarching operating system for the common VNF management tasks.

We illustrate below a simple method for an initial placement of the VNFs across the available servers. Given the available servers, the objective is to distribute the VNFs on them such that the load on each server is within its processing capacity and the total traffic across the servers is minimized. Since inter-server traffic incurs a higher delay penalty than the intra-server traffic, it is desirable to minimize the inter-server traffic. This problem can be viewed as a graph partitioning problem, which is known to be an NP-Hard combinatorial optimization problem. This method illustrated below is based on the Kernighan-Lin heuristic procedure for graph partitioning. Although this heuristics can lead to a "local optimal" solution that depends on the initial placement, it works well for practical purposes.

As shown in Figure 12.9, we have two service demands in this example. A service demand is a request for supporting a particular traffic class. Each service demand is specified by a linear graph showing the sequence of VNFs used along with the corresponding expected processing load as well as the data rate required between a pair of VNFs. For example, Figure 12.9 shows that Demand 1 requires VNFs F_A, F_B, and F_C in that order with the respective expected processing load of 40, 20, and 80 units, respectively, and expected data rates of 10 and 20 units from F_A to F_B, and F_B to F_C, respectively. The aggregate demand graph shows the sum of demands

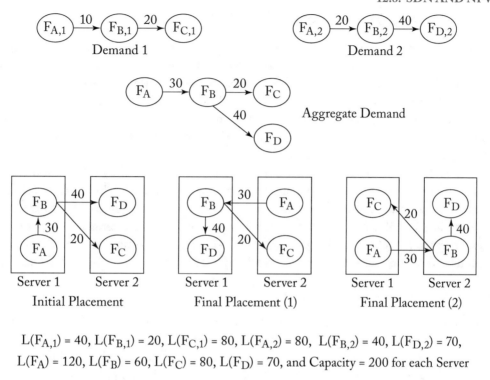

$L(F_{A,1}) = 40$, $L(F_{B,1}) = 20$, $L(F_{C,1}) = 80$, $L(F_{A,2}) = 80$, $L(F_{B,2}) = 40$, $L(F_{D,2}) = 70$,
$L(F_A) = 120$, $L(F_B) = 60$, $L(F_C) = 80$, $L(F_D) = 70$, and Capacity = 200 for each Server

Figure 12.9: VNF placement using Kernighan-Lin heuristics.

in terms of the expected processing load for the VNFs and the expected data rate between the VNFs. In this example, the total processing capacity at each server is 200 units. The initial placement shown in the figure is an arbitrary placement that satisfied the server processing capacity constraint. Feasibility of such initial placement is assumed as is the case for this example. Starting with the initial placement, the Kernighan-Lin heuristic procedure works as follows. In each iteration, consider one of the VNFs placed on one of the servers (say, Server 1), and then consider either just moving that VNF to the other server or swapping that VNF with another VNF on the other server. It is a valid move or swap if the processor capacity constraints would still be satisfied after the change. Consider all possible valid moves and swaps, and for this iteration, settle on the change that results in the smallest total inter-server data rate. In the next iteration, consider another VNF on Server 1, and repeat the procedure. At the end of these iterations, we find two equivalent final placements as shown in the figure. This heuristic procedure can be generalized to scenarios with more than two servers or scenarios where an aggregated VNF cannot be accommodated on one server.

12.9   INTERNET OF THINGS (IOT)

The IoT framework extends the set of users using the Internet to include a machine that may not even be under active human control, e.g., household and industrial machines, wearable devices, cars, sensors, etc. As such, the ideas of Sensor Networks and Machine-to-Machine (M2M) Communication are subsumed by this framework. The confluence of recent advances in the area of energy-efficient and miniaturized designs of the required hardware in the IoT endpoints, in networking technologies (especially, the wireless technologies), and in computation and storage server technologies is the key enabler for the IoT framework.

The explosion of the IoT endpoints poses many challenges. They include limits of addressing for the endpoints, traffic-carrying capacities of links and switching capacities of network nodes, QoS, and security. Since IoT endpoints can be range-limited due to the energy constraints, and can also have network access limitations due to interference from other such devices, energy management and network-access architecture (including specification of access algorithms and routing schemes) are additional important challenges for the IoT framework.

12.9.1   REMOTE COMPUTING AND STORAGE PARADIGMS

- Cloud Computing: Use a bank of servers hosted at a remote site (referred to as a Data Center) for processing and storing data. This is the first paradigm that moved computing and storage away from the hitherto common practice of using local servers for these purposes. Typically, the remote servers provide a very powerful computational platform with a large storage capacity.

- Fog Computing: This term was introduced by Cisco for referring to the paradigm that extends cloud computing to the network edge. Servers that may be co-located with the edge nodes, such as routers, provide a distributed environment for satisfying the computational and storage needs of IoT endpoints. This paradigm alleviates the ever expanding demands on the data centers, and can provide better performance by avoiding the need for data to travel to the centralized cloud sites.

- Mist Computing: This term refers to the paradigm that further distributes the computational and storage functions closer to the IoT endpoints by adding the required capabilities in the first-touch edge nodes like residential gateways and base stations. The basic idea is to strategically use the Mist, Fog, and Cloud Computing platforms in a hierarchical fashion, i.e., go only as far from the endpoints for satisfying the computational and storage requirements as necessary for an efficient implementation with the ability to provide good performance to the IoT endpoints.

In order to further help overcome the scalability issues in mobile networks, the proposed EdgeIoT framework requires each base station to be connected to mist and/or fog nodes for processing of data at the mobile edge. Furthermore, this framework calls for an SDN-based cellular core for forwarding data flows among mist and fog nodes as the prevailing loading conditions dictate. For the basic ideas behind SDN, please refer to the related section earlier in this chapter.

12.10 SUMMARY

This chapter reviews a few of the many additional topics related to networks.

- An *Overlay Network* is built on top of another network. Examples include peer-to-peer networks such as BitTorrent and content distribution networks such as Akamai. Two main issues are how to find resources and routing.

- A *wireless sensor network* is a collection of nodes equipped with sensors and radio transceivers. The nodes communicate in an ad hoc way. Issues include low-energy MAC protocols, location, routing, and reliability.

- Routers in the Internet and nodes in ad hoc networks implement distributed-routing protocols. Hosts in various networks may also implement various distributed algorithms. We discussed convergence properties of distributed applications, such as Bellman-Ford, TCP, and power control.

- We discussed the impossibility results for Byzantine agreement problems.

- Source compression enables economical storage and delivery of files, audio, and video streams. The chapter explains the *Lempel–Ziv* algorithm for file compression.

- SDN separates data and control planes. SDN and NFV can allow service providers to reduce their expenses while providing flexibility of introducing new services in a cost efficient and timely manner.

- IoT allows a machine not even under active human control to be an Internet endpoint. This poses a number of scalability issues. In order to address these issues hierarchical computational and storage paradigms are introduced: Cloud, Fog, and Mist Computing.

12.11 PROBLEMS

P12.1 This problem[2] illustrates the impact of Head-of-Line (HOL) blocking. Consider a 2X2 switch, and assume that there are always fixed sized packets available at the two input queues. The output ports the packets are destined for are independently and randomly chosen from the output ports 1 and 2 with equal probabilities. Let $d_i(n)$, $i = 1, 2$, denote the destination of the head-of-line packet of input queue i at time epoch n. If the two head-of-line packets at the two input queues are destined for different output ports at a

[2]Adapted from [96].

given epoch, both are transferred; otherwise one of the two packets is randomly chosen for transfer, while the other is blocked at its input queue.

(a) Let $d(n)$ denote the composition $d_1(n)d_2(n)$ (i.e., 11, 12, 21, or 22). Note that $d(n)$ is a Discrete Time Markov Chain (DTMC) with the state space 11, 12, 21, 22. Give the transition probability matrix of this DTMC, and compute its invariant distribution.

(b) Find the average throughput of this 2X2 switch.

P12.2 This problem[2] considers a simple maximal matching scheduler for a 2X2 VOQ switch. Assume fixed-size packets and that all input and output ports are operating at an identical link rate. The transmission time of a packet at this rate is referred to as a slot. Let $q_{ij}(n)$ denote the queue length at VOQ(i, j) (Virtual Output Queue at input i for output j) at time n. For $n > 0$, it coincides with the end of slot n. The arrival process into VOQ(i, j) is Bernoulli with mean λ_{ij}. The arrival processes are independent across queues and time slots. Assume $\lambda_{12} = 0$ and that all VOQs are empty initially. The scheduler works as follows for $n \geq 0$.

- If $q_{11}(n) > 0$ and $q_{22}(n) > 0$, a packet from VOQ(1, 1) and a packet from VOQ(2, 2) are scheduled.
- If $q_{11}(n) > 0$ and $q_{22}(n) = 0$, a packet from VOQ(1, 1) is scheduled.
- If $q_{11}(n) = 0$ and $q_{22}(n) > 0$, a packet from VOQ(2, 2) is scheduled.
- If $q_{11}(n) = 0$, $q_{22}(n) = 0$, and $q_{21}(n) > 0$, a packet from VOQ(2, 1) is scheduled.

(a) Why is this a maximal matching scheduler?

(b) Describe why you expect the maximum weight scheduler (MaxWeight scheduler) to be more complex to implement.

(c) Given $\lambda_{11} = \lambda_{22} = 0.5$, find the set of values for λ_{21} that can be supported by this scheduler.

(d) Instead of the scheduler described above, now consider the MaxWeight scheduler. Compare the set of supportable values of λ_{21} under the MaxWeight scheduler to that found in (c) above.

P12.3 Consider a 3X3 router as shown in Figure 12.10. The three output ports are labeled A, B, C. Time is slotted. Packets arrive at the three input buffers in the beginning of each slot. Each packet is labeled with the output port it must be forwarded to. At the end of each slot the fabric forwards packets to the output buffers. Packets are drained from the output buffers at a rate of one packet per slot.

For slots 1 through 5, the packet arrivals are labeled as follows and suppose there are no arrivals after slot 5.

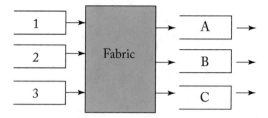

Figure 12.10: Figure for additional topics Problem 3.

- At input port 1: A, B, C, A, B.
- At input port 2: A, B, C, A, B.
- At input port 3: A, C, B, A, C.

(a) Compute the time at which all packets have departed the system when the router is input queued. Show your work. Break all ties by using the preference ordering $1 > 2 > 3$, (e.g., if two packets want to go to the same output port, forward the one from 1 before the one from 2).

(b) Repeat part (a) for when the speedup is 2 (i.e., the fabric can route at most two packets to the same output port in one slot time while taking at most one packet from each input port).

(c) Now suppose there are Virtual Output Buffers (VOBs) for each of the output ports at each of the input ports, and the fabric uses the MaxWeight Scheduler. Compute the time it takes for all packets to leave the system.

12.12 REFERENCES

Switches are discussed in [103]. The scheduling of crossbar switches is analyzed in [67], [65], and [96]. Fat trees are presented in [9]. For an overview of overlay networks, see [11]. Distributed algorithms are analyzed in [62]. See also [18] for recent results and [16] for a nice survey of some key issues. A cute animation of the Lempel-Ziv algorithm can be found at [12]. The distributed power control algorithm is from [33]. For a discussion on use of the SDN architecture for enterprises and service provider networks, see [64] and [24]. For an overview of OpenFlow, see [66]. [79] provides further details of the E2 framework. [19] discusses Fog Computing as the paradigm well suited to address the challenges posed by IoT. The EdgeIot framework is proposed in [99].

Bibliography

[1] 3GPP Specificaton Series 36. http://www.3gpp.org/ftp/Specs/html-info/36-series.htm 156

[2] 3GPP TR 38.913, Study on Scenarios and Requirements for Next Generation Access Technologies, 2016. 156

[3] 3GPP TS 23.203 V9.0.0, Policy and Charging Control Architecture (Release 9), March 2009. 156

[4] 3GPP TS 36.211 V8.6.0, Physical Channels and Modulation (Release 8), March 2009. 156

[5] 3GPP TS 36.300 V8.8.0, Overall Description, Stage 2 (Release 8), March 2009. 156

[6] N. Abramson and F. F. Kuo (Eds.), *Computer Communication Networks*, Prentice-Hall, 1973. 46

[7] Y. Abu-Mostafa, Learning from Data, http://work.caltech.edu/telecourse.html, 2012. 200

[8] R. Ahlswede, N. Cai, S. R. Li, and R. W. Yeung, Network information flow, *IEEE Transactions on Information Theory*, 2000. DOI: 10.1109/18.850663. 85

[9] M. Al-Fares, A. Loukissas, and A. Vahdat, A scalable, commodity data center network architecture, *Sigcomm*, 2008. DOI: 10.1145/1402958.1402967. 205

[10] M. Allman et al., TCP Congestion Control, RFC 2581, 1999. http://www.ietf.org/rfc/rfc2581.txt DOI: 10.17487/rfc2581. 114

[11] D. G. Andersen, Overlay networks: Networking on top of the network, *Computing Reviews*, 2004. http://www.reviews.com/hottopic/hottopic_essay_01.cfm 205

[12] Animation of the Lempel-Ziv algorithm. *Data-Compression.com*. http://www.data-compression.com/lempelziv.html 205

[13] V. Banos-Gonzalez et al., Throughput and range characterization of IEEE 802.11ah, *arXiv: 1604.08625v1* [cs.NI], April 28, 2016. DOI: 10.1109/tla.2017.8015044. 65

[14] P. Baran, On distributed communications networks, *IEEE Transactions on Communications Systems*, March 1964. `http://ieeexplore.ieee.org/search/wrapper.jsp?arnumber=1088883` DOI: 10.1109/tcom.1964.1088883. 7

[15] R. Bellman, On a routing problem, *Quarterly of Applied Mathematics*, 16(1), 87–90, 1958. DOI: 10.1090/qam/102435. 85

[16] D. P. Bertsekas and J. N. Tsitsiklis, Some aspects of parallel and distributed iterative algorithms—a Survey, *IFAC/IMACS Symposium on Distributed Intelligence Systems*, 1988. `http://www.mit.edu:8001/people/dimitrib/Some_Aspec` DOI: 10.1016/0005-1098(91)90003-k. 205

[17] G. Bianchi, Performance analysis of the IEEE 802.11 distributed coordination function, *IEEE Journal on Selected Areas in Communications*, 18(3), March 2000. DOI: 10.1109/49.840210. 53, 54, 56, 57, 65

[18] V. D. Blondel, J. M. Hendrickx, A. Olshevsky, and J. N. Tsitsiklis, Convergence in multiagent coordination, consensus, and flocking, *Proc. of the Joint 44th IEEE Conference on Decision and Control and European Control Conference*, 2005. DOI: 10.1109/cdc.2005.1582620. 205

[19] F. Bonomi, R. Milito, P. Natarajan, and J. Zhu, Fog computing: A platform for internet of things and analytics, *Big Data and Internet of Things: A Roadmap for Smart Environments*, Springer, 2014. DOI: 10.1007/978-3-319-05029-4_7. 205

[20] S. Boyd and L. Vandenberghe, *Convex Optimization*, Cambridge University Press, 2004. DOI: 10.1017/cbo9780511804441. 136, 140

[21] R. Braden (Ed.), Requirements for Internet Hosts—Communication Layers, RFC 1122. `http://tools.ietf.org/html/rfc1122` DOI: 10.17487/rfc1122. 28

[22] A. Cabellos et al., Knowledge-defined networking, *arXiv: 1606.06222v2*, [cs.NI], June 2016. DOI: 10.1145/3138808.3138810. 199

[23] D. Camps-Mur et al., Device-to-device communications with Wi-Fi direct: Overview and experimentation, *IEEE Wireless Communications*, 20(3), 96–104, 2013. DOI: 10.1109/mwc.2013.6549288. 65

[24] M. Casado et al., Fabric: A retrospective in evolving SDN, *ACM HotSDN*, 2012. DOI: 10.1145/2342441.2342459. 200, 205

[25] D.-M. Chiu and R. Jain, Networks with a connectionless network layer; part III: Analysis of the increase and decrease algorithms, *Technical Report DEC-TR-509*, Digital Equipment Corporation, Stanford, CA, August 1987. 7, 114

[26] D. Chiu and R. Jain, Analysis of the increase/decrease algorithms for congestion avoidance in computer networks, *Journal of Computer Networks and ISDN*, 17(1), 1–14, June 1989. http://www.cse.wustl.edu/~jain/papers/ftp/cong_av.pdf DOI: 10.1016/0169-7552(89)90019-6. 7, 114

[27] T. Clausen et al., Optimized Link State Routing Protocol (OLSR), RFC 3626, 2003. http://www.ietf.org/rfc/rfc3626.txt DOI: 10.17487/rfc3626. 81, 85

[28] G. Di Caro and M. Dorigo, AntNet: Distributed stigmergetic control for communications networks, *Journal of Artificial Intelligence Research*, 1998. 85

[29] E. W. Dijkstra, A note on two problems in connexion with graphs, *Numerische Mathematik*, 1, S. 269–271, 1959. DOI: 10.1007/bf01386390. 85

[30] R. Droms, Dynamic Host Configuration Protocol, RFC 2131, 1997. http://www.ietf.org/rfc/rfc2131.txt DOI: 10.17487/rfc2131. 95

[31] H. Ekstrom, QoS control in the 3GPP evolved packet system, *IEEE Communications Magazine*, 47(2), February 2009. DOI: 10.1109/mcom.2009.4785383. 156

[32] L. R. Ford and D. R. Fulkerson, Maximal flow through a network, *Canadian Journal of Mathematics*, 8, 399-404, 1956. DOI: 10.4153/cjm-1956-045-5. 116

[33] G. Foschini and Z. Miljanic, A simple distributed autonomous power control algorithm and its convergence, *IEEE Transactions on Vehicular Technology*, 1993. DOI: 10.1109/25.260747. 205

[34] V. Fuller, T. Li, J. Yu, and K. Varadhan, Classless Inter-Domain Routing (CIDR): an Address Assignment and Aggregation Strategy, RFC 1519, 1993. http://www.ietf.org/rfc/rfc1519.txt DOI: 10.17487/rfc1519. 7, 28

[35] M. Gast, *802.11 Wireless Networks: The Definitive Guide*, 2nd ed., O'Reilly, 2005. 51, 53, 65

[36] R. Gibbens and F. P. Kelly, Measurement-based connection admission control, *5th International Teletraffic Congress*, June 1997. 166

[37] T. G. Griffin and G. Wilfong, An analysis of BGP convergence properties, *SIGCOMM'99*. DOI: 10.1145/316194.316231. 85

[38] P. Hoel, S. Port, and C. Stone, *Introduction to Probability Theory*, Houghton-Mifflin Harcourt, 1977. 46

[39] P. Hoel, S. Port, and C. Stone, *Introduction to Stochastic Processes*, Houghton-Mifflin Harcourt, 1972. 65

[40] G. Huston et al., Textual Representation of Autonomous System (AS) Numbers, RFC 5396, 2008. http://www.ietf.org/rfc/rfc5396.txt DOI: 10.17487/rfc5396. 7

[41] IEEE 802.11. http://en.wikipedia.org/wiki/IEEE_802.11 % 41 65

[42] IEEE 802.11-1999 Specification. DOI: 10.1016/s1353-4858(00)80001-x. 51, 65

[43] IEEE 802.11a-1999 Specification. 53, 65

[44] IEEE 802.11ac-2013 Specification. 65

[45] IEEE 802.11ad-2012 Specification. 65

[46] IEEE 802.11b-1999 Specification. 53, 65

[47] IEEE 802.11g-2003 Specification. 65

[48] IEEE 802.11n-2009 Specification. 65

[49] Van Jacobson, Congestion avoidance and control, *SIGCOMM'88*. Later rewritten with M. Karels. http://ee.lbl.gov/papers/congavoid.pdf DOI: 10.1145/52324.52356. 7, 114

[50] Y. Ji et al., Average rate updating mechanism in proportional fair scheduler for HDR, *IEEE Globecom*, 3464–3466, 2004. DOI: 10.1109/glocom.2004.1379010. 156

[51] L. Jiang and J. Walrand, A distributed CSMA algorithm for throughput and utility maximization in wireless networks, *Allerton*, 2008. DOI: 10.1109/allerton.2008.4797741. 140

[52] B. Kahn and V. Cerf, A protocol for packet network intercommunication, *IEEE Transactions of Communications Technology*, May 1974. DOI: 10.1109/TCOM.1974.1092259. 7

[53] S. Katti, H. Rahul, W. Hu, D. Katabi, M. Medard, and J. Crowcroft, XORs in the air: Practical wireless network coding, *SIGCOMM'06*, Pisa, Italy, September 11–15, 2006. DOI: 10.1109/tnet.2008.923722. 85

[54] F. P. Kelly, A. Maulloo, and D. Tan, Rate control for communication networks: Shadow prices, proportional fairness and stability, *Journal of the Operational Research Society*, 49(3), 237–252, 1998. DOI: 10.1038/sj.jors.2600523. 140

[55] E. Khorov et al., A survey of IEEE 802.11ah: An enabling networking technology for smart cities, *Computer Communications*, 2014. DOI: 10.1016/j.comcom.2014.08.008. 65

[56] L. Kleinrock, Message Delay in Communication Nets with Storage, Ph.D. thesis, Cambridge, Massachusetts Institute of Technology, 1964. http://dspace.mit.edu/bitstream/handle/1721.1/11562/33840535.pdf 7

[57] L. Kleinrock and R. Muntz, Multilevel processor-sharing queueing models for time-shared systems, *Proc. of the 6th International Teletraffic Congress*, Munich, Germany, 341/1–341/8, September 1970. 166

[58] I. Koutsopoulos and L. Tassiulas, Channel state-adaptive techniques for throughput enhancement in wireless broadband networks, *IEEE Infocom*, 2, 757–766, April 2001. DOI: 10.1109/infcom.2001.916266. 156

[59] R. Liao et al., MU-MIMO MAC protocols for wireless local area networks: A survey, *arXiv: 1404.1622v2* [cs.NI], November 28, 2014. DOI: 10.1109/comst.2014.2377373.

[60] J. D. C Little, A proof of the Queueing Formula $L = \lambda W$, *Operations Research*, 9, 383–387, 1961. DOI: 10.1287/opre.9.3.383. 28

[61] S. H. Low and D. E. Lapsley, Optimization flow control, I: Basic algorithm and convergence, *IEEE/ACM Transactions on Networking*, 1999. DOI: 10.1109/90.811451. 140

[62] N. Lynch, *Distributed Algorithms*, Morgan Kaufmann, 1996. 205

[63] Max-Flow Algorithms. https://en.wikipedia.org/wiki/Maximum_flow_problem % 63 116

[64] N. McKeown, Software-defined networking, *IEEE Infocom*, April 2009. 205

[65] N. McKeown, The iSLIP scheduling algorithm for input-queued switches, *IEEE/ACM Transactions on Networking (TON) archive*, 7(2), 188–201, April 1999. DOI: 10.1109/90.769767. 205

[66] N. McKeown et al., OpenFlow: Enabling innovation in campus networks, *ACM Sigcomm*, April 2008. DOI: 10.1145/1355734.1355746. 205

[67] N. McKeown, A. Mekkittikul, V. Anantharam, and J. Walrand, Achieving 100% throughput in an input-queued switch, *IEEE Transactions on Communications*, 47(8), 1260–1267, 1999. DOI: 10.1109/26.780463. 205

[68] P. Maillé, S. Parekh, and J. Walrand, Overlaying delay-tolerant service using SDN, *Networking*, 2016. DOI: 10.1109/ifipnetworking.2016.7497227. 198, 199

[69] M. Mathis et al., TCP Selective Acknowledgment Options, RFC 2018, 1996. http://www.ietf.org/rfc/rfc2018.txt DOI: 10.17487/rfc2018. 102

[70] R. M. Metcalfe and D. R. Boggs, Ethernet: Distributed packet switching for local computer networks, Xerox Parc Report CSL757, May 1975, reprinted February 1980. A version of this paper appeared in *Communications of the ACM*, 19(7), July 1976. http://ethernethistory.typepad.com/papers/EthernetPaper.pdf DOI: 10.1145/357980.358015. 46

[71] J. Mo and J. Walrand, Fair end-to-end window-based congestion control, *IEEE/ACM Transactions on Networking* 8(5), 556–567, October 2000. DOI: 10.1109/90.879343. 140

[72] P. Mockapetris, Domain Names—Concepts and Facilities, RFC 1034, 1987. `http://www.ietf.org/rfc/rfc1034.txt` DOI: 10.17487/rfc1034. 28

[73] J. Mogul and J. Postel, Internet Standard Subnetting Procedure, RFC 950, 1985. `http://www.ietf.org/rfc/rfc950.txt` DOI: 10.17487/rfc0950. 95

[74] A. Molisch, *Wireless Communications*, 2nd ed., Wiley, 2011. 156

[75] H. Myung, J. Lim, and D. Goodman, Single carrier FDMA for uplink wireless transmission, *IEEE Vehicular Technology Magazine*, September 2006. DOI: 10.1109/mvt.2006.307304. 156

[76] J. Nagle, On packet switches with infinite storage, *IEEE Transactions on Communications*, 35(4), April 1987. DOI: 10.1109/tcom.1987.1096782. 166

[77] M. J. Neely, E. Modiano, and C-P. Li, Fairness and optimal stochastic control for heterogeneous networks, *Proc. of IEEE Infocom*, 2005. DOI: 10.1109/infcom.2005.1498453. 140

[78] W. B. Norton, Internet Service Providers and Peering, 2000. `http://www.cs.ucsd.edu/classes/wi01/cse222/papers/norton-isp-draft00.pdf` 85

[79] S. Palkar et al., E2: A framework for NFV applications, *ACM Symposium on Operating Systems Principles*, October 2015. DOI: 10.1145/2815400.2815423. 205

[80] A. K. Parekh and R. G. Gallager, A generalized processor sharing approach to flow control in integrated service networks : The single node case, *IEEE/ACM Transactions on Networking*, 1(3), June 1993. DOI: 10.1109/90.234856. 166

[81] C. Perkins et al., Ad hoc On-Demand Distance Vector (AODV) Routing, RFC 3561, 2003. `http://www.ietf.org/rfc/rfc3561.txt` DOI: 10.17487/rfc3561. 80, 85

[82] D. C. Plummer, An Ethernet Address Resolution Protocol, RFC 826, 1982. `http://www.ietf.org/rfc/rfc826.txt` DOI: 10.17487/RFC0826. 95

[83] J. Postel (Ed.), Transmission Control Protocol,' RFC 793, 1981. `http://www.ietf.org/rfc/rfc0793.txt` DOI: 10.17487/rfc0793. 114

[84] J. Proakis, *Digital Communications*, McGraw-Hill, 2000. 177

[85] A. Puri and S. Tripakis, Algorithms for routing with multiple constraints, In *AIPS'02 Workshop on Planning and Scheduling using Multiple Criteria*, 2002. 85

[86] R. Ramaswami and K. N. Sivarajan, *Optical Networks—A Practical Perspective*, 2nd ed., Morgan Kauffman, 2000. 177

[87] Recommendation ITU-R M.2083-0, IMT Vision—Framework and Overall Objectives of the Future Development of IMT for 2020 and Beyond, September 2015. 156

[88] J. Saltzer, D. Reed, and D. D. Clark, End-to-end arguments in system design, *2nd International Conference on Distributed Computing Systems*, 509–512, April 1981, *ACM Transactions on Computer Systems*, 2(4), 277–288, November 1984. DOI: 10.1145/357401.357402. 28

[89] Y. Rekhter et al., A Border Gateway Protocol 4 (BGP-4), RFC 1771, 1995. http://www.ietf.org/rfc/rfc1771.txt DOI: 10.17487/rfc4271. 85

[90] G. Romano, 3GPP RAN Progress on 5G, 3GPP Presentation, 2016. 156

[91] N. Ruangchaijatupon and J. Yusheng, Simple proportional fairness scheduling for OFDMA frame-based wireless systems, *IEEE Wireless Communications and Networking Conference*, 1593–97, 2008. DOI: 10.1109/wcnc.2008.285. 156

[92] C. E. Shannon, A mathematical theory of communication, *Bell System Technical Journal*, 27, 379–423/623–656, 1948. http://cm.bell-labs.com/cm/ms/what/shannonday/shannon1948.pdf DOI: 10.1002/j.1538-7305.1948.tb00917.x. 28

[93] A. Shokrollahi, Raptor codes, *IEEE Transactions on Information Theory*, 52, 2551–2567, 2006. DOI: 10.1109/tit.2006.874390. 85

[94] T. Slater, Queuing Theory Tutor. http://www.dcs.ed.ac.uk/home/jeh/Simjava/queueing/ 28

[95] R. Srikant, *The Mathematics of Internet Congestion Control*, Birkhäuser, 2004. DOI: 10.1007/978-0-8176-8216-3. 140

[96] R. Srikant and L. Ying, *Communication Networks: An Optimization, Control, and Stochastic Networks Perspective*, Cambridge University Press, 2013. 140, 156, 203, 205

[97] I. Stojmenovic, Position based routing in ad hoc networks, *IEEE Communications Magazine*, 40(7):128–134, 2002. DOI: 10.1109/mcom.2002.1018018. 85

[98] W. Sun et al., IEEE 802.11ah: A long range 802.11 WLAN at sub 1 GHz, *Journal of ICT Standardization*, 1, 83–108, 2013. 65

[99] X. Sun and N. Ansari, EdgeIoT: Mobile edge computing for the internet of things, *IEEE Communication Magazine*, 54(12), December 2016. DOI: 10.1109/mcom.2016.1600492cm. 205

[100] L. Tassiulas and A. Ephremides, Stability properties of constrained queueing systems and scheduling policies for maximum throughput in multihop radio networks, *IEEE Transactions on Automatic Control*, 37, 1936–1948, 1992. DOI: 10.1109/cdc.1990.204000. 140

[101] D. Tse and P. Viswanath, *Fundamentals of Wireless Communication*, Cambridge University Press, 2005. DOI: 10.1017/cbo9780511807213. 156

[102] G. Tsirtsis et al., Network Address Translation—Protocol Translation (NAT-PT), RFC 2766, 2000. http://www.ietf.org/rfc/rfc2766.txt DOI: 10.17487/rfc2766. 95

[103] P. Varaiya and J. Walrand, *High-performance Communication Networks*, 2nd ed., Morgan Kaufman, 2000. 205

[104] J. Walrand, *Probability in Electrical Engineering and Computer Science: An Application-driven Course*, Amazon, 2014. 65

[105] J. Wannstorm, LTE-Advanced, 3GPP, June 2013. 156

[106] Wi-Fi Direct. http://www.wi-fi.org/discover-wi-fi/wi-fi-direct 65

[107] H. Zhu and R. Hafez, Scheduling schemes for multimedia service in wireless OFDM systems, *IEEE Wireless Communications*, 14, 99–105, October 2007. DOI: 10.1109/mwc.2007.4396949. 156

Authors' Biographies

JEAN WALRAND

Jean Walrand received his Ph.D. in EECS from UC Berkeley, and has been on the faculty of that department since 1982. He is the author of *An Introduction to Queueing Networks* (Prentice Hall, 1988), *Communication Networks: A First Course* (2nd ed., McGraw-Hill, 1998), and *Probability in Electrical Engineering and Computer Science* (Amazon, 2014), and co-author of *High-Performance Communication Networks* (2nd ed., Morgan Kaufman, 2000), *Scheduling and Congestion Control for Communication and Processing Networks* (Morgan & Claypool, 2010), and *Sharing Network Resources* (Morgan & Claypool, 2014). His research interests include stochastic processes, queuing theory, communication networks, game theory, and the economics of the Internet. Prof. Walrand is a Fellow of the Belgian American Education Foundation and of the IEEE, and a recipient of the Informs Lanchester Prize, the IEEE Stephen O. Rice Prize, the IEEE Kobayashi Award, and the ACM Sigmetrics Achievement Award.

SHYAM PAREKH

Shyam Parekh received his Ph.D. in EECS from UC Berkeley in 1986. He is currently an Associate Adjunct Professor in the EECS department at UC Berkeley. He has previously worked at AT&T Labs, Bell Labs, TeraBlaze, and ConSentry Networks. He was a co-chair of the Application Working Group of the WiMAX Forum during 2008. He is a co-editor of *Quality of Service Architectures for Wireless Networks* (Information Science Reference, 2010). He currently holds 10 U.S. patents. His research interests include architecture, modeling, and analysis of both wired and wireless networks.

Index